THE SICAMOUS & THE NARAMATA

Steamboat Days in the Okanagan

Robert D. Turner

CONTENTS

A PROJECT OF
THE S.S. SICAMOUS SOCIETY
IN CO-OPERATION WITH THE
ROYAL BRITISH COLUMBIA MUSEUM

PUBLISHED BY
Sono Nis Press
VICTORIA, BRITISH COLUMBIA

Canadian Cataloguing in Publication Data

Turner, Robert D., 1947-
The Sicamous and the Naramata

Includes bibliographical references.
ISBN 1-55039-057-0

1. Sicamous (Paddle steamer) 2. Naramata (Steamboat) 3. Paddle steamers—British Columbia—Okanagan Valley—History. 4. Steam-navigation—British Columbia—Okanagan Valley—History. I. Title.
HE566.P3T87 1995 386′.22436′09711 C95-910367-8

The SS Sicamous Society in co-operation with the Royal British Columbia Museum, gratefully acknowledges financial assistance for this project from the British Columbia Ministry of Small Business, Tourism and Culture.

Publication costs assisted by the Canada Council Block Grant Program

A project of
THE SS SICAMOUS SOCIETY
in co-operation with the
ROYAL BRITISH COLUMBIA MUSEUM

Published by
SONO NIS PRESS
1745 Blanshard Street
Victoria, British Columbia
V8W 2J8

Designed and printed in Canada by
MORRISS PRINTING COMPANY LTD.
Victoria, British Columbia

Acknowledgements

The preparation of this book was made possible because of the help and encouragement of many people. I would like to thank: Edward L. Affleck; Clement Battye; A. H. "Barney" Bent; Richard E. Brown; Jim and Anthea Bryan; Peter Corley-Smith; Maurice Chandler; Bill Curran; Bonnie Dafoe; the late George Donaldson, chief engineer; Lillian Estabrooks; Hugh Fraser; Captain Bill Guttridge, first officer; Bill Hillyard, ship's carpenter; Doug Jones; W. Gibson Kennedy; Dr. Herb McGregor; Ruth Hansen McGregor; Fred McKie, chief engineer; Captain Robert Manning, first officer; Robert F. Marriage; Earl Marsh; the late Bill Merrifield, ship's carpenter; Mary Merrifield, relief cook; Glen Morley; L. S. Morrison; Captain Norman Nordstrom; the late Mary Orr; Milton Parent; Robert W. Parkinson; Captain Sam Podmoroff; William Ramsay, deckhand and shipyard worker; Ron Roberts, second engineer; Captain Ed Schimpf; the late Eric Sismey; Bob Spearing; the late Captain Walter Spiller; the late Bill Triggs, purser; the late Charles Verey, chief engineer; Ray Vinten; the late Ed Vipond; Ed Wanstall, shipyard worker and deckhand; the late David Webster, chief steward; Dave Wilkie; Austin Willet, deckhand; the late John Williams, chief engineer; Brian Wilson, Heritage Photo Co-op; and Wayne Wilson. Many people who worked on the sternwheelers and tugs had qualifications for senior positions but, because of seniority requirements and the shortage of jobs, had to work in other jobs. The credit lines in the quotations do not necessarily reflect the position of the individuals at the time of the incident being told.

The members of the SS Sicamous Restoration Society and the Kettle Valley Railway Heritage Society have made great efforts to see the *Sicamous* and the *Naramata* preserved and they have made this project a pleasure. Thanks in particular to: Diane Bagno; Hartley Clelland; Doug Griffith, site manager; Larry Little; Ian MacLeod; Randy Manuel, who is also director of the Penticton Museum; Jack Petley, who was assistant superintendent of the CPR's Kettle Valley Division; Marlene Pugh; David Stocks; and Fred Tayler. Similarly, my sincere thanks to the Royal British Columbia Museum, in particular Jim Wardrop, and the Ministry of Small Business, Tourism and Culture for their assistance with the publication of this book.

Edward Affleck, Peter Corley-Smith, Martin Lynch, Randy Manuel and Nancy Turner all read the manuscript and I sincerely appreciate their

"To me the *Nasookin*, the *Bonnington* and the *Sicamous* were floating palaces for their time."—CHIEF ENGINEER JOHN WILLIAMS, 1979

"When I was outside, I would spend most of my time watching the paddlewheel and those huge engines turning the wheel around. It's a percussive sort of trip; the faster she would go, the more you would feel the blow. It was two solids hitting one another; it's noisy."—GLEN MORLEY, 1994

"I remember a moonlight cruise, which was kind of a farewell party when they knew it was going to be taken off completely. That must have been about 1935. It was a beautiful night and crowded, well-patronized. There wasn't dinner, just buffet type of refreshments. They danced. I was torn between the beautiful moonlight on deck and wanting to watch the passing of the landscape in the moonlight and the fact that a sing-song started around the piano inside. Mrs. Harold Glass could play by ear anything anybody sang. I think the music won. I spent most of the evening in at the piano."—BONNIE DAFOE, 1993

The *Sicamous* and *Naramata* at Penticton during the heyday of the Okanagan steamboats. Two vessels could hardly appear less alike but their careers paralleled each other from their construction in 1914 through to their preservation at Penticton.
—LUMB STOCKS, HERITAGE PHOTO CO-OP

thoughtful suggestions, insights and comments. Christopher Seton assisted with research in Canadian Pacific Archives.

In addition, thanks go to the following organizations and their ever-co-operative staff and volunteers for information and photographs: Arrow Lakes Historical Society, Nakusp; British Columbia Archives and Records Service (BCARS), Victoria; British Columbia Legislative Library, Victoria; British Columbia Orchard Industry Museum, Kelowna; Canadian Pacific Archives, Montreal; Kelowna Centennial Museum; Kootenay Lake Historical Society (Kootenay Lake Archives) and the SS *Moyie* National Historic Site, Kaslo; National Archives of Canada, Ottawa; Nelson Museum (Kootenay Museum Association and Historical Society); Penticton Museum; Summerland Museum; and the Greater Vernon Museum and Archives.

This book draws on many sources including contemporary reports in Okanagan newspapers, recollections and diaries, interviews carried out over the past 20 years with many who worked on or travelled on the sternwheelers and tugs, company correspondence, steamship inspection records, and photographs. In addition, the vessels themselves have revealed many aspects of their history through the workmanship, technology and ambience they preserve.

The Canadian Pacific Railway's monogram was etched into the plate glass doors leading to the dining saloon.
—ROBERT D. TURNER

"The meals were those wonderful CPR meals. You couldn't beat them anywhere. The people working on them were very proud of their boats and of their jobs. They had a good rapport up and down the lake. People depended on them.
—GLEN MORLEY, 1994

"They served all the meals in the CPR style, with the male stewards with the napkins over their arms and all the array of silver, cutlery and dishes."
—BONNIE DAFOE, 1993

The *Sicamous* at Kelowna on July 23, 1937, decorated with flags for what proved to be her last excursion. It was sponsored by the Penticton Gyro Club.
—KELOWNA MUSEUM

The dining room or dining saloon of the *Sicamous* was the showpiece of the vessel. Located on the saloon deck with the surrounding gallery in the deck above, it provided seating for 50 to 70 patrons.
—CHARLES VEREY COLLECTION

Naramata's career was spent in comparative obscurity but the tug played a key role moving the Okanagan's prized fruit to market for over 50 years.
—VERNON MUSEUM & ARCHIVES

"It seemed as if the entire population of the town turned out to see the soldiers. . . . As the soldiers had to live on their regular rations for the day they greatly enjoyed a meal served them on the boat with the dessert consisting of 500 pounds of cherries. As the boat pulled out for the north the soldiers gave hearty cheers, replied to by the town crowd on the wharf."
—*The Penticton Herald*, JULY 13, 1916

The *Sicamous* was the setting for many emotional farewells but few could match the departure of the troops from Okanagan communities for training camps, England and eventually the trenches of the Western Front during the First World War.
—PENTICTON MUSEUM

Introduction

"It was five days by train from Ottawa and we stayed overnight, of course as everybody did, at Sicamous at the CPR hotel. And then we started down the valley by train as far as Vernon and we went out to Okanagan Landing and transferred onto the *Sicamous*. That was the usual transportation up and down the valley."
—BONNIE DAFOE, 1993

"On Monday . . . , 2,100 crates of cherries were shipped out on the *Sicamous* by Dominion Express. General daily shipments are running from 1,100 to 1,500 crates a day . . . an output of 10,000 crates a week is not by any means an item to be sneezed at."
—*The Kelowna Courier*, JULY 29, 1920

The history of steam navigation on British Columbia's lakes and rivers has many facets, and marks a period of enormous changes in the way people lived, worked and travelled. The steamboats, mostly sternwheelers, operated in many conditions and parts of the province. They could run through the swift, treacherous rapids of the Skeena or the Upper Fraser, or steam almost silently across the frequently tranquil deep waters of Okanagan Lake or Kootenay Lake. In the Okanagan, the careers of the steamers were marked by contrasts: the elegance and romance of moonlight cruises and the hectic pace of the late summer and early fall fruit rush; the toil of working under scorching summer heat or battling miles of ice jams when the winds could chill to the bone. Wherever they ran, these lake and river boats were fascinating and their role was central to the people and the communities they served.

This is the story of two very different Okanagan steam vessels, the *Sicamous* and the *Naramata*: the first a luxurious paddlewheeler, the second a workaday tug. The *Sicamous*, like many other sternwheelers, was a multi-purpose vessel designed to combine elaborate, first-class passenger accommodations with a large cargo capacity. The shallow-draught hull and sternwheel propulsion permitted beach landings, safe operations in water depths of just a few feet and stops at small communities without extensive wharves. A large sternwheeler like the *Sicamous* could slip into a dock in water so shallow that a coastal steamer of equivalent size and capacity would have run hard aground. The simple and functional propulsion system was also easy to maintain, and comparatively inexpensive to build. The sternwheelers were fast compared to tugs and barges, which in the Okanagan helped them move the competitive, highly valuable and perishable fresh fruits to market. The tugs, like *Naramata*, had a different role. They were built to handle barges efficiently and reliably between permanent docks and transfer slips along the lake routes. They were much smaller, slower vessels and provided little or no passenger accommodation or facilities for mail and express, but in the end they would be the more enduring.

Despite their obvious differences, they have much in common. Both were operated by the Canadian Pacific Railway on Okanagan Lake and both were completed in 1914 just before the beginning of the First World War. Their jobs depended largely on the success and production of the fruit-growing communities around Okanagan Lake in the busy years of the early 1900s. Both, in different ways, became key links in the transportation system of the district and important parts of the social and economic fabric of the communities they served. Even in technology, they had features in common: they came from the same shipyard and they had steel hulls and modern engines. Finally, both have been preserved at Penticton on the shore of the lake where they spent their working lives. Both are important and precious heritage landmarks that will continue to make substantial cultural and economic contributions to the Okanagan for future generations.

The *Sicamous*, much larger and having carried thousands of people, has become symbolic of the region's history while the less glamorous *Naramata* has only recently emerged from obscurity to take its place as a very important heritage vessel. The *Sicamous* operated for just over 23 years while the *Naramata* went about its less exciting duties for over half a century. Their story has deeper roots that reach back to the 1890s when steam navigation was just beginning in the Okanagan and it continues after their retirement to include the efforts by many dedicated people to see the two steamers preserved for future generations to enjoy. The *Sicamous* and the *Naramata* have many stories to tell; this book presents some of them along with glimpses of the other Okanagan Lake steamers, with the help of memories from people who worked and travelled on them.

"Four hundred happy pleasure seekers, the largest excursion crowd the SS *Okanagan* has carried for some time, made the trip to Kelowna on Thursday to attend the annual picnic of the combined Penticton Sunday Schools. A day of unalloyed pleasure, not marred with even the most minor accident, was enjoyed by the crowd that packed themselves into the big lake steamer."—*The Penticton Herald*, JULY 18, 1914

"Squally Point was probably the worst. Cross winds. Lots of times, we'd go in behind the island and wait. The old-timers before me, they steamed there and didn't make any headway; they were going backwards. The *Naramata* was only 125 brake horsepower, not very much. Two loaded barges with 16 cars on them."
—CAPTAIN NORMAN NORDSTROM, 1994

From *The Lake District of Southern British Columbia*, published by the CPR in 1919.
—ROBERT W. PARKINSON COLLECTION

The Aberdeen and the S&O Railway

When the Canadian Pacific Railway was completed across southern British Columbia in November 1885 it penetrated a huge area divided by enormous mountain barriers and cut by rivers and lakes. To the south of the CPR main line was a region that stretched a couple of hundred miles to the United States border and included, potentially, some of the finest farming, grazing and orchard land in Canada. There were several important areas, but one of the largest was the Okanagan Valley, part of the Columbia River drainage. There were some large areas of flat lands but often the pockets of fertile land were small, situated on deltas of side streams or on the dry benches above the lakeshore. Agricultural development, mining or settlement needed a transportation system that could give the farmers and growers or others access to both supplies and markets.

The Shuswap & Okanagan Railway was chartered and incorporated in 1886 to build a line from Sicamous, on the CPR main line, to the head of Okanagan Lake near Vernon. The CPR leased the railway and in 1891 construction began. The next year tracks reached Vernon and were extended to Okanagan Landing on the lakeshore about five miles (eight km) to the southwest. South of Vernon, railway construction would have been prohibitively expensive but, at the same time, the southern parts of the valley were slowly becoming settled and could someday produce an abundance of fruit. The answer was a reliable steamer service down the lake. Initially primitive but gradually improving services had been provided for some time, the most famous operator being Captain Thomas D. Shorts who faithfully plied the lake for several years with his sail-assisted rowboat, the *Ruth Shorts*, named for his mother. His later and more sophisticated vessels were the *Mary Victoria Greenhow*, the *Jubilee*, the *City of Vernon* and the *Penticton*, which was acquired later by the Lequime brothers who operated a sawmill at Kelowna, and then passed to other owners.

The need for a modern steamer was clear to CPR management, and President William C. Van Horne authorized construction in 1892. The CPR sought the advice of Captain James W. Troup of Nelson who was manager of the Columbia & Kootenay Steam Navigation Company (C&KSN), which operated steamer services on the Columbia River and Kootenay Lake in connection with the CPR. Considered the leading authority on steamboat

"The time generally taken for the round voyage was nine days, and wherever night overtook them the *Ruth Shorts* was pulled ashore and the crew camped for the night.

"'When will we get there?'

"'Haven't the slightest idea, but rest assured that we'll fetch up there some time.'"—CAPTAIN THOMAS SHORTS, QUOTED IN *The Vernon News*, JUNE 15, 1893

design in British Columbia, he toured the area and recommended a substantial vessel that would handle the anticipated traffic from the district for some years. A few years later, when the CPR bought the C&KSN's operations to form what became known as the British Columbia Lake & River Service, Troup was appointed manager. Later he moved to Victoria to manage the CPR's expansion into coastal shipping but he continued to be called upon in matters of vessel design and for major decisions involving the Lake & River Service. He was a particularly gifted designer of vessels and his style was carried through the Lake & River Service fleet as well as the coastal *Princess* liners. Troup was succeeded by Captain John Clancy Gore, a capable and experienced officer who, like Troup, was also a skilled steamboat designer.

The new vessel for the Okanagan was named the *Aberdeen*, after John Campbell Gordon, 7th Earl of Aberdeen. He was the owner of the large Coldstream Ranch at Vernon and was appointed Governor-General of Canada in 1893. A classic western sternwheeler, and described as "the finest inland steamer set afloat in the Northwest in 1893,"[1] it was 146 feet (44.5 m) in length, with a cargo deck capacity of about 200 tons (180 tonnes) and spacious and pleasant passenger accommodations, which included 11 staterooms, a dining room, smoking room and ladies' saloon. Designed by John F. Steffen (or Steffin), a Danish shipbuilder from Portland, the steamer was constructed under the supervision of Edwin G. McKay, master builder, who also came from Oregon. Engines were supplied by the B.C. Iron Works in Vancouver, to the designs of Horace Campbell, and the CPR itself built the boiler in Montreal. A shipyard to build and maintain the new steamer was established at Okangan Landing. Captain J. Foster was the first master, R. Williams was mate and W. Couson (or Cousens) the chief engineer. When the *Aberdeen* went into service in June 1893, it was just as if the CPR had opened a branch line extending all the way from Okanagan Landing to Penticton. People living in communities down the valley had direct, reliable, comfortable access to the CPR main line. For the CPR, the *Aberdeen* was an investment in the future potential of the Okanagan. It would take over a decade for the fruit industry and many communities to become well established.

The *Aberdeen* settled into a routine of service to the communities along Okanagan Lake, starting a pattern that would endure until the mid-1930s. At first infrequent, the service soon expanded to provide three return trips each week between Okanagan Landing and Penticton. Winter service was suspended from mid-January 1894 to mid-March and the steamer *Penticton*

CANADIAN PACIFIC RAILWAY.

PACIFIC DIVISION.

SHUSWAP & OKANAGAN BRANCH.

TIME TABLE NO. 7, TO TAKE EFFECT ONE O'CLOC

JUNE 15TH, 1892.

TRAINS SOUTH. READ DOWN.				TRAINS NORTH. READ UP.	
Okanagan Mixed **No. 14**	Miles from Sicamous Jnctn.	Telegraph Calls.	STATIONS.	Miles from Okanagan Lndg.	Shuswap Mixed **No. 13**
8.05	0.	R S	De. **Sicamous Junction.** Ar§	51.0	**18.50**
			12.9		
*8.44	12.9		.... MARA	38.1	18.00*
			10.5		
9.15	23.4	D	............ENDERBY.........§	27.6	17.30
			8.3		
9.45	31.7	M S	.........ARMSTRONG	19.3	17.00
			6.3		
*10.05	38.0		.... LARKIN	13.0	16.40*
			8.1		
a 10.30 d 10.45	46.1	N O	.... VERNON ....§	4.9	d 16.15 a 16.00
			4.9		
11.00	51.0	K	Ar. **Okanagan Landing.** De.	0.	**15.45**
Okanagan Mixed **No. 14**	Miles from Sicamous Jnctn.	Telegraph Calls.	STATIONS.	Miles from Okanagan Lndg.	Shuswap Mixed **No. 13**

*FLAG STATION. §WATER TANK.

☞ TRAINS RUN DAILY EXCEPT SUNDAY.

—CANADIAN PACIFIC ARCHIVES

"The President [William C. Van Horne] decided on his last visit to Vernon that we should put a steamer on Okanagan Lake, and if we can extend the route to the foot of Dog Lake [Skaha Lake] I think it would be very desirable to do so, as it would open up that much more country and make it tributary to the Shuswap Railway."—GENERAL SUPERINTENDENT HARRY ABBOTT TO VICE-PRESIDENT T. G. SHAUGHNESSY, JUNE 10, 1892

[1] E. W. Wright, 1895, *Lewis & Dryden's Marine History of the Pacific Northwest*, p. 408. (Reprinted in 1967 by Superior Publishing Co., Seattle.)

The Shuswap & Okanagan Railway, built between Sicamous on the main line of the Canadian Pacific and Okanagan Landing, connected the Okanagan Valley with the transcontinental railway. The railway opened to Vernon in June 1892.
—VERNON MUSEUM & ARCHIVES

The completion of the steamer *Aberdeen* in 1893 extended CPR service the length of Okanagan Lake as far south as Penticton. Like the construction of most railway branch lines, the establishment of the steamer service by the CPR was an investment in the future and a necessary step if the area was to be settled and developed for agriculture, lumbering or mining.—KELOWNA MUSEUM

"I proposed to employ Captain Troup, who is universally recognized as the best steamboat authority on the Coast, to examine the route and the country, and to report as to the dimensions and class of steamer he would recommend as the most suitable for the business. The next proposed step was to employ a regular builder to furnish a specification and model based on the general figures given by Captain Troup. This has all been done [and] the specifications furnished by Mr. Steffen, who is one of the most experienced builders of steamers of this class on the [Puget] Sound."

—HARRY ABBOTT TO
T. G. SHAUGHNESSY, SEPTEMBER 8, 1892

The *Aberdeen*, shown at Okanagan Landing in 1902 in one of a pair of overlapping photos, was a western sternwheeler of classic design. Her shallow-draught hull and paddlewheel enabled her to pull into shallows with little danger of grounding. Passengers enjoyed the amenities of a large saloon deck with a ladies' saloon, smoking lounge, dining room and overnight staterooms, while there was ample capacity for freight. The *Aberdeen* ran until 1916 and was sold in 1919 and eventually broken up.

—BULMAN PHOTO,
HERITAGE PHOTO CO-OP

OKANAGAN LAKE (PENTICTON) ROUTE

STEAMER ABERDEEN

Monday, Wednesday, Friday........	10.30	........OKANAGAN LANDING........	13.00	...Tuesday, Thursday and Saturday
	14.00	..............KELOWNA.............	[illegible].30	
	15.30	............PEACHLAND............	8.00	
	17.30	.............PENTICTON............	6.00	

Okanagan Landing, just five miles (eight km) southwest of Vernon, developed into a bustling community as the interchange point between the steamers and the S&O Railway. As traffic expanded with the growth in population in the valley, the rail yards were enlarged, facilities built for icing refrigerator cars, and a transfer slip for railcar barges was constructed.—BULMAN PHOTO, HERITAGE PHOTO CO-OP

The officers and crew of the *Aberdeen* pose in front of the pilothouse in 1898. In the front row, left to right, are: J. McDonald, purser; Joe Weeks, mate; Captain George L. Estabrooks; R. Hawes, chief engineer; and E. Petman, steward. Behind them are a waiter called "Slabs"; a deckhand; A. McDonald, freight clerk; A. Finlayson; and W. A. Tip, pantryboy. In the back row are W. Grasse and W. Gibbs, at left; and Tom Jones, second from right; the other deckhands are unknown.—C. W. HOLLIDAY, HERITAGE PHOTO CO-OP

Lambly's, now called Peachland, shown in 1902, right, was typical of many steamer landings in the Okanagan. Small docks, warehouses and perhaps a store with post office were the centres of these communities. Originally the steamers would have simply pulled in as close to shore as possible and the deckhands would have slid out a gangplank. They stopped just long enough to drop off and pick up passengers, freight, mail and express. Often the captain would hold the vessel against the dock by keeping the paddlewheel turning slowly. In fruit season, the deckhands would be busy trying to get the fruit on board without undue delay to the vessel's already tight schedule.
—BULMAN PHOTO, HERITAGE PHOTO CO-OP

Ice plagued the crews and maintaining even very elastic winter schedules sometimes became impossible. Off Penticton in thick ice, the *Aberdeen* pushes a barge to help clear a channel to the dock. Residents are out on the ice to watch and help chop a way for the steamer. The wooden-hulled steamers could be badly damaged by the thick ice sheets.—KELOWNA MUSEUM

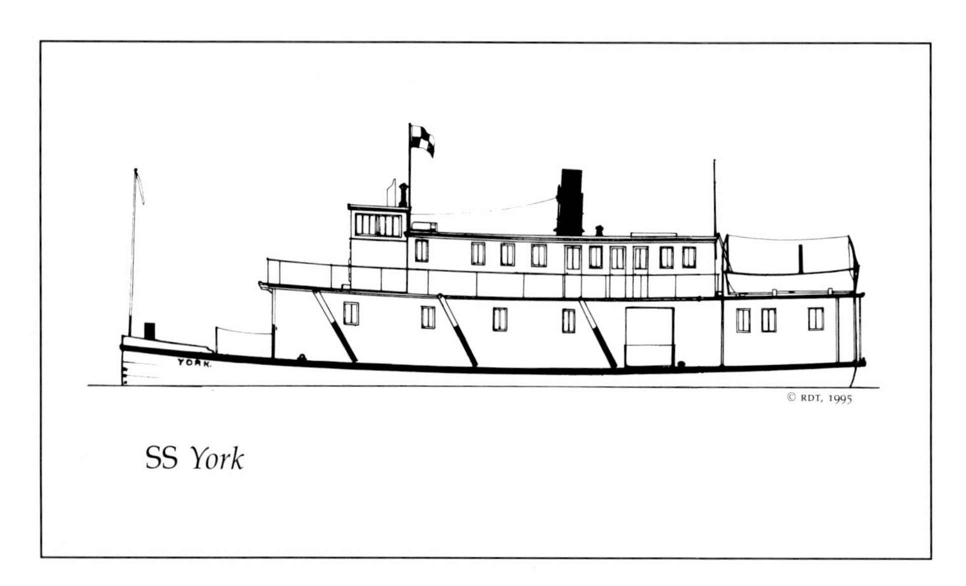

SS *York*

The Steamer York

The *York*, a trim vessel of 134 gross tons, was to operate for nearly 30 years in the Okanagan. By 1931 when no longer needed by the CPR, it had steamed nearly 300,000 miles (480,000 km). Sold to mill operator Sid Leary at Nakusp on the Arrow Lakes, the *York* was never returned to service.

The *York* pushes an ice-breaking barge, with cargo, to clear a channel for the *Aberdeen* in February 1907. Captain Joe Weeks was on the *York* while Captain George L. Estabrooks was in command of the *Aberdeen*.—KELOWNA MUSEUM

Too small and slow to substitute effectively for the *Okanagan* or *Sicamous* and lacking in enough power to be an efficient tug on Okanagan Lake, the *York* nonetheless proved to be a useful addition to the fleet during busy times and in service to the smaller communities. The modest proportions were not a particular disadvantage on Skaha, or Dog, Lake during the 1920s when it was only necessary to move one barge.
—BCARS, HP57206

"They would signal if there was a bride on board . . . they'd blow four blasts and half the town would come down to see the new bride. Mother came in here [to Kelowna] as a bride in 1905 on the *Aberdeen*. They used to have their ways of letting the public know. . . . Meet them at the gangplank and wish them well."
—BILL KNOWLES, 1994

"I am told that business on Okanagan Lake is increasing very much, in fact, it is very nearly up to the capacity of the 'Aberdeen' and it is expected that this season it will be beyond her capacity under present conditions. . . ."
—CAPTAIN JAMES W. TROUP, APRIL 18, 1904

filled in on a twice-weekly schedule. In later years the steamers *Fairview* and *Greenwood* filled in for the *Aberdeen*. Until the 1907 construction of the sternwheeler *Okanagan*, the *Aberdeen* normally ran southbound on Mondays, Wednesdays and Fridays, leaving Okanagan Landing in mid-morning and arriving at Penticton in late afternoon. Returning north, it left Penticton early in the morning on Tuesdays, Thursdays and Saturdays and arrived back at the Landing early in the afternoon. Cargo loading, unexpected stops to pick up passengers along the way, or unusually heavy traffic could delay the schedule considerably.

In many places wharves were very primitive or non-existent and there were few amenities for travellers, but improvements were made and business for the *Aberdeen* grew. Mining developments in the south Okanagan and adjacent areas added large volumes of traffic during the 1890s. The agricultural economy of the Okanagan took time to develop; irrigation often was required, orchards had to mature and markets were needed. By the early 1900s the work of the last decade and the increasing numbers of people moving to the region was having its effect. Substantial crops began to be harvested and the export of fruit grew, establishing the Okanagan's reputation for fine cherries, apricots, peaches, plums, apples and pears. Later, cantaloupe, zucca melon and other crops were added and all contributed to the traffic on the steamers.

The CPR was not alone in providing service on Okanagan Lake and over the years there were privately owned tugs, launches and small passenger vessels, such as those operated by the Okanagan Lake Boat Company, that provided a variety of local services. By the early 1900s, the CPR clearly needed an additional vessel on the lake to handle growing traffic and provide a relief steamer for the *Aberdeen*. This was the small, propeller-driven *York*, built with a seven-section, prefabricated steel hull and two compound engines by the Bertram Engine Works in Ontario. The steamer was assembled at Okanagan Landing and launched on January 18, 1902. Apparently originally intended for use on Trout Lake in the West Kootenay, the *York*, with a licence for 90 passengers, filled in as a tug, a service for which it was not designed, and passenger steamer on Okanagan Lake and was later used on Skaha, or Dog, Lake to the south of Penticton during the 1920s. There it connected a Kettle Valley Railway line between Okanagan Falls and Haynes, south of Oliver, with the tracks to Penticton. In 1931 trackage was completed along Skaha Lake, which made the steamer service unnecessary, and in 1944 the branch line was extended south as far as Osoyoos.

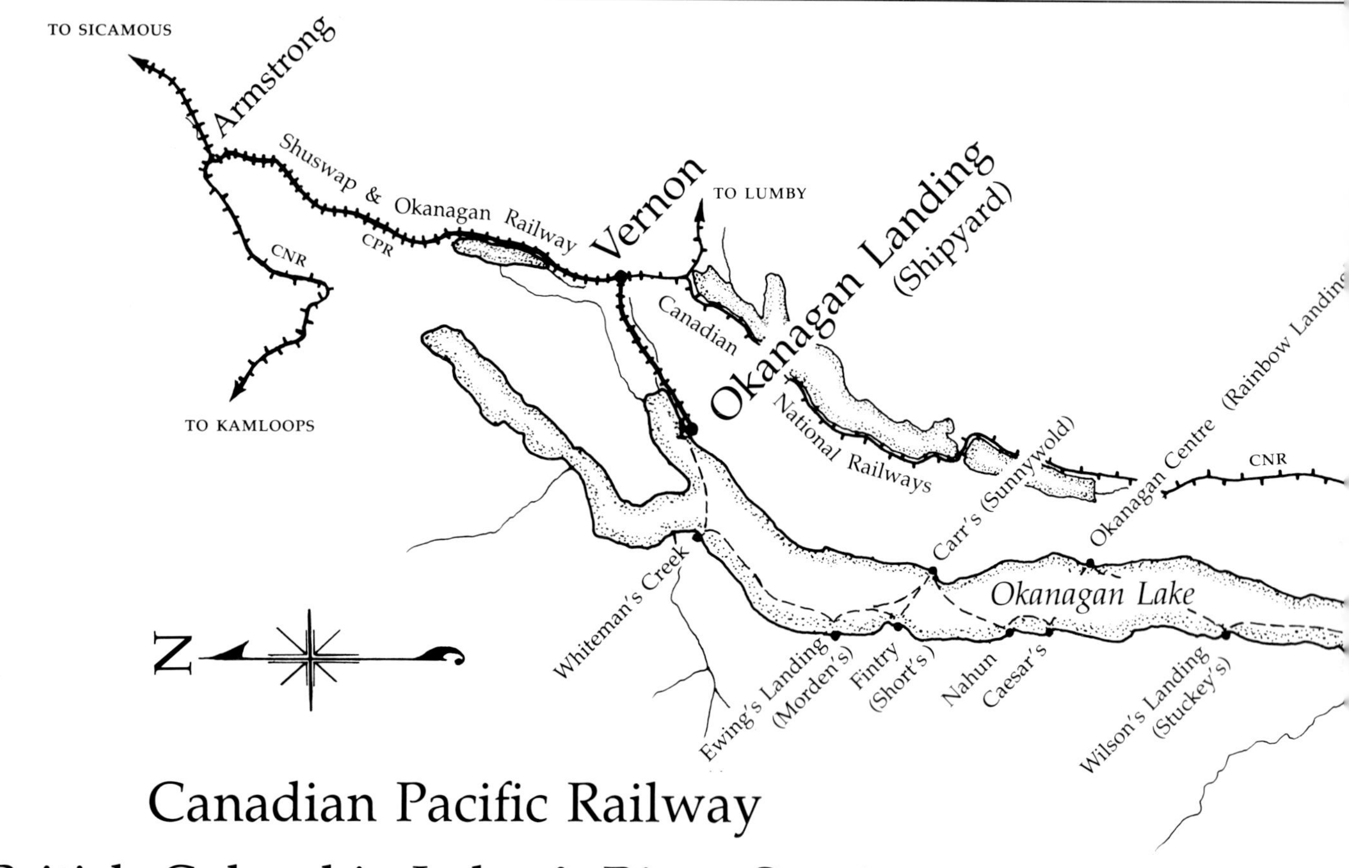

Canadian Pacific Railway
British Columbia Lake & River Service
Okanagan Services, 1893-1972

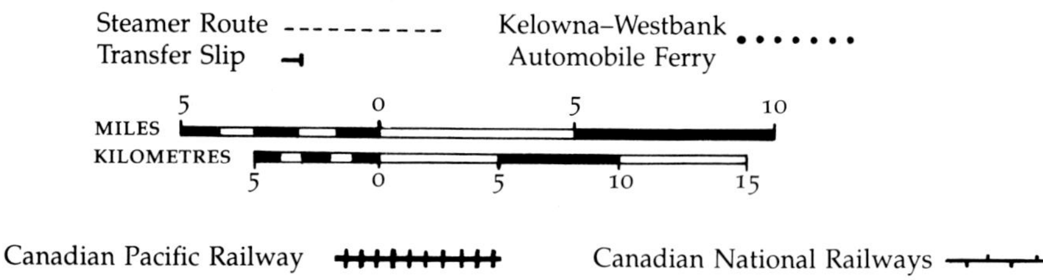

Okanagan Lake Steamer Route, Distances

Okanagan Landing*		0 Miles
Shorts	9.5 miles	(15 km)
Nahun	13.5 miles	(22 km)
Rainbow Landing	16.5 miles	(27 km)
Kelowna*	28.5 miles	(46 km)
Gellatly	36.5 miles	(59 km)
Peachland	43 miles	(69 km)
Greata	49.5 miles	(80 km)
Summerland*	56.5 miles	(90 km)
Naramata†	56 miles	(90 km)
Penticton*	64.5 miles	(104 km)

*location of CPR transfer slip
†CNR transfer slip used by CPR
(approximate distance)

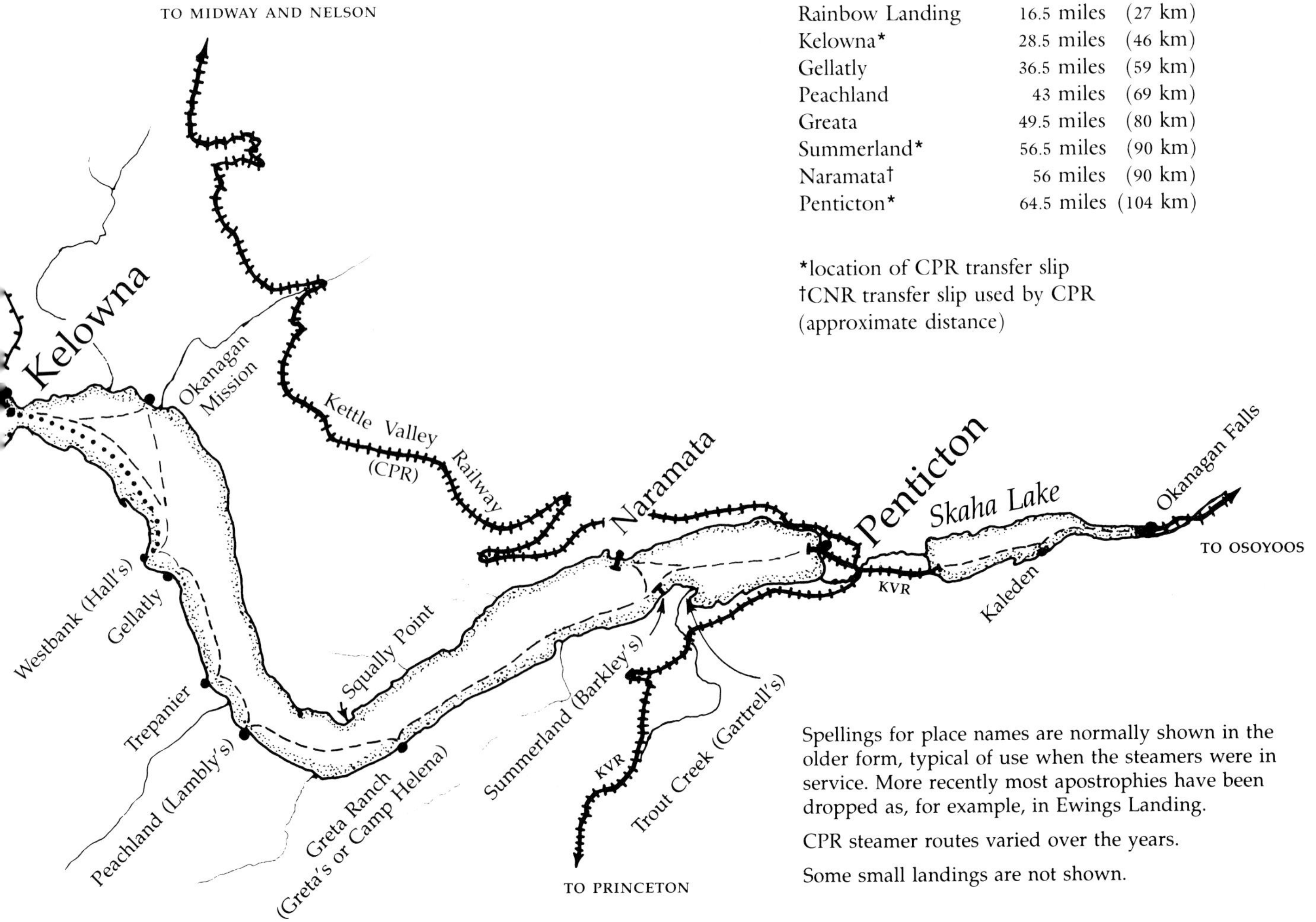

Spellings for place names are normally shown in the older form, typical of use when the steamers were in service. More recently most apostrophies have been dropped as, for example, in Ewings Landing.

CPR steamer routes varied over the years.

Some small landings are not shown.

OKANAG

The Okanagan

"At 3:20, everything now being in readiness, Mr. Bulger [the CPR's master builder], who was ashore directing affairs, gave the signal, and the eight men, each with a broad axe, as one man, severed the eight ropes and in little more than 30 seconds, the noble ship glided down the ways, amidst the sound of hundreds of cheering voices on board and ashore, with the steamers whistling, the locomotive tooting and bells ringing, the boat took to the water like the proverbial duck."
—*The Okanagan Semi-Weekly*, APRIL 16, 1907

The launching of the beautiful sternwheeler *Okanagan* on April 16, 1907, was celebrated with a ball at Okanagan Landing's Strand Hotel, hosted by Captain and Mrs. Gore. The orchestra was under the direction of W. R. Rainey.
—PROBABLY A G. H. E. HUDSON PHOTO, HERITAGE PHOTO CO-OP, OPPOSITE; VERNON MUSEUM & ARCHIVES, BELOW

By 1904, it was becoming clear that traffic on Okanagan Lake was growing beyond the capacity of the ageing *Aberdeen* and the CPR began developing plans for a new vessel. A similar situation was apparent on Kootenay Lake, where the elegant steamer *Moyie*, built in 1898, was proving too small for the important service between Kootenay Landing and Nelson. Captain Troup, Captain Gore and other senior CPR officials recommended construction of steel-hulled sternwheelers but, in the end, wooden-hulled vessels, initially cheaper but more expensive in the long run because of higher maintenance costs, were authorized for both services. They were developed from the proven design of the beautiful express steamer *Rossland*, built for the company's Columbia River service in 1897. The first, the *Kuskanook*, was completed at Nelson in 1906 and the new boat for Okanagan Lake followed. Appropriately, it was to be named *Okanagan*. The new steamers were described as three-deckers: the first for freight and machinery; the second (the saloon deck) included a bar, men's smoking room, dining saloon, ladies' lounge, offices and overnight cabins; while the third or upper deck provided more staterooms, an observation lounge and officers' quarters. Above these decks was the pilothouse.

Preliminary work began on the *Okanagan* in the spring of 1906 and the keel was laid on July 10 at the Okanagan Landing shipyard. Through the remainder of 1906 and into the spring of 1907, the CPR's crew of highly skilled tradesmen worked on the complex task of building the 193-foot-long (58.8 m) steamer. On April 16, 1907, the work was nearly done. A special train brought guests and the Vernon Band to the Landing for launching ceremonies. To the cheers of a large crowd and the music of the band, "Mrs. Gore," noted *The Okanagan Semi-Weekly*, "broke the traditional beribboned bottle of champagne over the bows saying: 'I call her *Okanagan*; success to her.' The beautiful boat then floated out on the bosom of Okanagan Lake as graceful as a swan. All on board now repaired to the dining saloon where the waiters served claret cup and everyone drank to the steamer *Okanagan* and the Okanagan district." Newspaper reports noted the cost at approximately $90,000.

Completed later that spring, the *Okanagan* brought a dramatic improvement in service to the communities along Okanagan Lake. Faster and more luxurious than the *Aberdeen* by a substantial margin, she also freed the older steamer to be used for freight and improved services to smaller settlements around the lake. That summer a daily steamer schedule was established with two trains a day connecting Vernon to the CPR main line at Sicamous.

The pace of settlement and transportation developments in southern British Columbia was accelerating and during the next few years many even more dramatic improvements would be made. By the early 1900s, railways had been proposed to just about every corner of southern British Columbia but still only the S&O actually reached Okanagan Lake. However, subsidiaries of the Great Northern Railway were building north from Washington into the Similkameen mining camps and the CPR was expanding across southern B.C. By the summer of 1910, the CPR was beginning the construction of the Kettle Valley Railway connecting the end of its trackage in the Boundary District with its main line across the Fraser River from Hope. Competition with the Great Northern was often intense, and both companies constructed trackage to many points along the southern boundary of the province before practicality and the First World War forced compromises and agreements. The Kettle Valley Railway was completed between Penticton and Midway to the east in 1914 and was opened through the Coquihalla Pass and to the CPR main line in 1916.

Kettle Valley Railway construction had an immediate effect on the steamer services as workers and supplies started arriving in the Okanagan. Penticton became a major centre for the construction crews working both east and west and an expanded barge service was needed. However, before the construction placed major demands on the Okanagan fleet, the small sternwheeler *Kaleden* was completed in August 1910 for service on the Okanagan River and Skaha Lake south of Penticton. The high hopes were frustrated; the river proved difficult to navigate despite dredging and work on the channel, so the *Kaleden* was used as an extra freight boat and tug on Okanagan Lake.

Although barges could be, and were, handled by sternwheelers on Okanagan Lake, deep-hulled, propeller-driven tugs were a far more suitable type of vessel for moving large, often heavily laden, barges. To fill this need, handle increasing construction traffic for the Kettle Valley and improve service to the growing fruit industry, the large wooden-hulled, steam tug *Castlegar*, was launched at Okanagan Landing on April 12, 1911. This handsome steamer was similar to the CPR's successful screw-driven tugs used in the Kootenays. A barge slip was built at Penticton in 1911 to permit

The *Okanagan* made several trial and excursion trips in April and May 1907 before entering regular service that June. —G. H. E. HUDSON, KELOWNA MUSEUM

"The new CPR steamer *Okanagan* in a trial trip last Sunday made the run between Penticton and Vernon's port [Okanagan Landing], a distance of 65 miles [105 km], in three hours and fifteen minutes, or at the rate of over 21 miles [34 km] per hour. She's a greyhound. . . ."—*The Okanagan Semi-Weekly*, APRIL 23, 1907

The *Okanagan* and her sistership *Kuskanook* were two of the most beautiful lake and river steamers in western North America.—VERNON MUSEUM & ARCHIVES

Summerland, north of Penticton, became one of the busiest fruit shipping centres in the Okanagan. By 1910, fruit volumes had reached levels requiring a tug and barge service, and transfer slips were eventually built at the larger communities. The new steam tug *Castlegar* has brought an eight-car barge to the Summerland transfer slip while the *Okanagan*, on its regular run, is at the dock in the distance.
—C. PEELE, HERITAGE PHOTO CO-OP

The *Castlegar* with two eight-car barges. The photo shows the way the tugs usually pushed two barges.
—G. H. E. HUDSON, KELOWNA MUSEUM

loaded railcars to be moved directly on barges from Okanagan Landing to the new railway.

The tugs and barges contributed substantially to the capacity of the steamer service on the lake and the fleet of deck barges and railcar transfer barges grew steadily. Over the years they carried thousands of railcars of fruit and produce from the Okanagan to markets across the continent. Two wooden deck barges were built in 1906 as part of the same appropriation as the *Okanagan*, and another for use in icebreaking was built in 1912. Eight-car wooden transfer barges were built in 1908, 1910, 1912 and 1913 and more followed after the end of the First World War. Additional barge slips were built eventually by the CPR at Summerland, Okanagan Centre, Kelowna, and Westbank. Later, the Canadian National built a slip at Naramata that the CPR also used and the CNR was permitted use of the slip at Summerland. In this way major packing houses could load produce directly into boxcars or refrigerator cars, which could be moved quickly to market.

Until the outbreak of the First World War, many people were moving to southern British Columbia and tourism was growing rapidly. To meet the growth in traffic that seemed inevitable, the CPR began a further expansion of its B.C. Lake & River Service. This period of growth and optimism brought about the construction of three of the largest and most modern sternwheelers ever to operate in British Columbia, and an expansion of tug and barge services.

The sternwheeler *Kaleden*, launched on July 23, 1910, was built to extend services to Kaleden and Okanagan Falls on Skaha or Dog Lake as it was then known. However, navigation on the Okanagan River was too difficult and service on Skaha Lake was left to smaller companies. The *Kaleden* was used as an extra boat on Okanagan Lake but was withdrawn from service during the First World War. —KELOWNA MUSEUM

Purser's stamp from the *Okanagan*.

The *Okanagan* breaks the perfect calm of an early morning on its way north from Penticton in 1907.—PROBABLY A C. PEELE PHOTO, HERITAGE PHOTO CO-OP

ICAMOUS

Building the Sicamous and the Naramata

"A hoarse roar, deeper than the voice of any other craft on the lake, heralded the approach of the fine new CPR steel steamer *Sicamous* at about 12:45 p.m. on Friday [June 12, 1914] at the local dock. Owing to the time being the dinner hour, there were few on the wharf to greet her, but her deep-throated announcement of her arrival speedily brought a large crowd from every portion of the town."
—THE KELOWNA *Courier*, JUNE 18, 1914

"I remember we were having dinner in the house. . . . And we heard this deep horn, this steam horn, which was different from the *Okanagan* and different from the *Aberdeen* so we knew our boat had arrived. I can see Dad yet, opening up the little window above the dining room table so we could hear it. And of course we all rushed down to watch her come in."
—BILL KNOWLES, IN 1994, REMEMBERING THE ARRIVAL OF THE *Sicamous* AT KELOWNA IN 1914

Towering over a proud shipyard crew and some family members, the *Sicamous* is nearly ready for its launching at the Okanagan Landing shipyard on May 19, 1914.—CHARLES VEREY COLLECTION

The first of this last generation of sternwheelers was the *Bonnington*, built for the Columbia River–Arrow Lakes service. This magnificent steamer was based on the plans proposed earlier by Captain Troup and Captain Gore when they had recommended steel steamers at the time of the construction of the *Kuskanook* and the *Okanagan*. Originally the *Bonnington* was to have a hull 193 feet (58.8 m) long, but Captain Troup suggested it be lengthened to 200 feet (61 m) to give the vessel more buoyancy and speed. The prefabricated hull, the boiler and engines were built by the Polson Iron Works in Toronto. Unlike the *Okanagan* and the other wooden-hulled vessels, the new steamers were to be four-deckers, with massive cabins that would dwarf anything afloat on the Interior lakes and rivers of British Columbia. In fact, they were some of the largest sternwheel steam vessels ever built for service in Western North America, and were exceeded in tonnage by only a few vessels, notably several on the Sacramento River, including the famous *Delta King* and *Delta Queen*, and in length by a few other west coast or northern steamers, such as the *Klondike*, now preserved as a National Historic Park at Whitehorse.

The plans for the new vessels combined many features of the *Kuskanook* and *Okanagan* with an earlier design highlight of the Columbia River steamers *Nakusp* and *Kootenay* built in the mid-1890s. The deck above the dining saloons of these vessels was open in the centre forming a gallery or balcony overlooking the dining saloon. This gave the saloons of the new steamers a spaciousness that was particularly pleasing, and "awesome," as Ted Affleck recalled, "to a small boy already overwhelmed by the menu."

Overall, there was a "family" appearance to nearly all of the CPR's sternwheelers. There were several variations that reflected design and operating experience, service conditions and the evolving architectural style of the team who produced a long series of exceptional vessels from the 1890s through the construction of the last new CPR sternwheelers just before the First World War. Key in this group of men were Captain James Troup; Captain John Gore; Thomas J. Bulger (master builder until his retirement in 1903) and his sons James and David, who both became master builders;

David Stevens, the Lake & River Service's senior engineer, who designed the hull and engines for the *Bonnington* and her sisterships, and George H. Keys, foreman joiner.

The *Bonnington*, completed at Nakusp in 1911, was followed in 1913 by the *Nasookin* for the busy Nelson-to-Kootenay Landing run. The new sternwheeler for the Okanagan was named *Sicamous* after the town and junction between the main line of the CPR and the Shuswap & Okanagan Railway at the eastern end of Shuswap Lake. The name was derived from the Shuswap language and meant "narrow" or "squeezed in the middle," referring to the narrow channel where Mara Lake empties into Shuswap Lake.

The design of the *Sicamous* followed on the successful pattern of the *Bonnington* and *Nasookin*, but with a shorter boat deck or Texas deck because of the anticipated need for fewer overnight cabins. The *Nasookin* and the *Sicamous* also differed from the *Bonnington* in hull design. The *Bonnington* was built with a flat-bottomed riverboat hull because of her route through the Columbia River narrows near Burton while the two later vessels had the more rounded hull form of a lake steamer. The contract for the hull, boiler and machinery was let to the Western Dry Dock & Ship Building Co. of Port Arthur (now Thunder Bay), Ontario. The hull, built of steel but with a wrought iron stem, was divided into 20 watertight compartments. The coal bunker was designed to have a watertight bottom to form a tank between the "skin [and] longitudinal and transverse bulkhead." The specifications noted that "great care [is] to be taken to have this tank thoroughly tight as this may be required for fuel oil." Although oil was cleaner burning, was much easier to load onto a vessel and did not require the labour of coal passers or stokers, the *Sicamous* was never converted. The hull was temporarily erected at Port Arthur, with the frames bolted together rather than riveted, and was then dismantled for shipment to Okanagan Landing. The engines were also assembled and tested before being shipped west.

A further addition to the growing Okanagan fleet was ordered from the Port Arthur yard on July 23, 1913. This was a modern, steel-hulled tug to expand the capacity of the tug and barge service. It was to be named *Naramata* after the prosperous fruit-growing community north of Penticton. Like the *Sicamous*, this vessel was to be prefabricated and shipped to Okanagan Landing for assembly but while the passenger steamer had wooden cabins, the tug was of all-steel construction except for finishing work. The tug was first assembled at Port Arthur and then "knocked down"

"The Hull to be built at Contractor's yard, taken down and shipped, re-erected and finished at Okanagan Landing, B.C. The Railway Company to furnish transportation for all men Material and Tools to Okanagan Landing, also return transportation for men and tools to Contractor's Yard, also to supply suitable building site also material for sufficient Blocking and Staging, free of expense to Contractors, for finishing the hull."
—SPECIFICATION FOR STEEL TUG FOR OKANAGAN LAKE, 1913

"When they built the *Sicamous*, my mother wanted to go up to the Landing and see it launched because she had watched some of her dad's sailing ships being built and launched [in Nova Scotia]. So she took me on the *Okanagan*. I was six. They launched her broadside. I guess nobody had taken the trouble to check the depth of the lake and she stuck in the mud, broadside. They hitched up a big rope to the bow of the *Okanagan* and tied it to the bow of the *Sicamous* and then they set the big paddlewheel going. I can still see this rope coming up tighter and tighter and tighter but they couldn't move her."—BILL KNOWLES, 1994

The yard was busy as the steel hulls of both *Sicamous* and *Naramata* took shape (top right) and later when the shipwrights and ship's carpenters began to frame the cabins (top left). This view shows the dining saloon at an early stage of construction. The steamer, ready for launching, dwarfed the passenger cars in the yard.—HERITAGE PHOTO CO-OP; BCARS, HP1419; EARL MARSH COLLECTION

for movement by rail to Okanagan Landing. The contract price was $26,500 but by the time the vessel was in operation, the cost was closer to $40,000. Having the riveting crew, which was brought to Okanagan Landing from Port Arthur, assemble two hulls saved a considerable amount of money.

Work began on the new steamers in September 1913, and through the winter of 1913-14 and into the following spring, the Okanagan Landing shipyard was alive with activity. With as many as 150 men employed in building the two steamers, there was a shortage of accommodation at Okanagan Landing and a morning and evening train was run between the Landing and Vernon for the workers. James M. Bulger, from Nelson, was master builder and George H. Keyes was the foreman joiner.

Supplies, materials and the prefabricated components for the two new vessels came in by the carload. Nineteen railway cars were required to ship the *Bonnington*'s hull and engines to Nakusp in 1910; the *Sicamous* probably required the same. As the warm days of early spring came to the Okanagan, the frames for the *Sicamous* were taking shape and the hull and cabins for the *Naramata* were nearing completion. The tug was finished first and it was launched on April 20, 1914.

With an all-steel hull, powerful compound engine and pleasing lines, the new steamer was an impressive addition to the small fleet of tugs on Okanagan Lake. Like the other tugs in the Lake & River Service, the *Naramata* was a utilitarian vessel with few frills. Although it nominally had a passenger capacity of 20 people, in its early years the *Naramata*'s assignment was almost invariably the barge service. The hull provided sleeping quarters for the crew in the bow, bunkers for coal, the boiler room and the engine room. The deckhouse included the galley with a small dining area for the crew, toilet, shower, and one cabin. The pilothouse was situated well forward and elevated to provide a good view ahead of the steamer. Behind the pilothouse were cabins for the captain and first officer. The new tug was attractively painted: a green hull with buff band and buff cabins. "Naramata," highlighted in gold leaf, was carefully applied on a black nameboard on the front of the pilothouse.

The *Naramata* normally carried a crew of the captain, one mate (first officer), a pilot during the busy fruit season, two deckhands, a chief engineer and a second engineer, two fireman, one bargeman and a cook, who also served as steward. On entering service, the *Naramata* took her place as the most modern tug on the southern lakes and rivers. While newer wooden-

Sicamous was launched at 2:15 p.m. on May 19, 1914.—PENTICTON MUSEUM

hulled tugs were built, using the engines and designs of retired tugs, the *Naramata* was not really outclassed until the 1928 construction of the *Granthall* for Kootenay Lake.

The *Sicamous* was launched just under a month later at 2:15 p.m., on Tuesday, May 19, 1914, and was christened by Mrs. J. I. E. Corbet, the daughter of Captain Gore and wife of CPR superintendent Corbet. "To the tooting of whistles and cheering of the large crowd . . . the splendid new steamer *Sicamous* slid gracefully into the water . . ." wrote the reporter for Vernon's *News*. But there she stopped. Apparently the launching cradles stuck in shallow water, the lake not being as high as hoped. Heavy lines were run onboard from the *Okanagan* and *Castlegar* but they could not pull the *Sicamous* free. The *Okanagan*, which had brought many people to the Landing to see the launching, had to return south but at 7:00 p.m. the *Castlegar* and the *Aberdeen*, working together, were able to float the big steamer. No one let the temporary difficulties mar the event, although it is certain there were worried moments for the shipyard crew and CPR officials. Mrs. Corbet was presented with an upholstered chair from the shipyard workers and a bouquet of carnations by the Misses Stobo and Reid, daughters, respectively, of shipwright John Stobo, who later became CPR master builder at Nelson, and Captain Matthew Reid of the *Castlegar*.

"The furnishings and fittings of the steamer," noted *The Vernon News* on May 21, 1914, "have been done in Australian mahogany and in teak wood from Burma, a combination which gives an effect of unusual richness. The large observation and smoking rooms back of the upper tier of staterooms is practically walled with plate glass and will afford a splendid point of vantage from which to view the scenery. . . . Writing desks and reading lamps will be installed and a piano is to be put on the balcony above the dining room. The most modern fire fighting devices will be provided. . . . In addition to the usual staterooms for the crew, there are special rooms for the mail clerk and express messenger, a cold storage room for meats and poultry, a pastry room, shower bath for the crew and other conveniences."

At the forward end of the saloon deck was the smoking room and at the aft end was the ladies' observation room. On the gallery deck or upper deck, above, the positions of the ladies' saloon and the smoking room were reversed. Along each side of the two passenger decks were the staterooms: 12 on the saloon deck and 24 on the upper deck. Cabins were numbered from one through 37; number 13 was omitted. There were four large stateroom's

The *Naramata* was launched on April 20, 1914.—PENTICTON MUSEUM

The new tug proved to be a reliable and sturdy vessel on the demanding barge service.—EARL MARSH COLLECTION

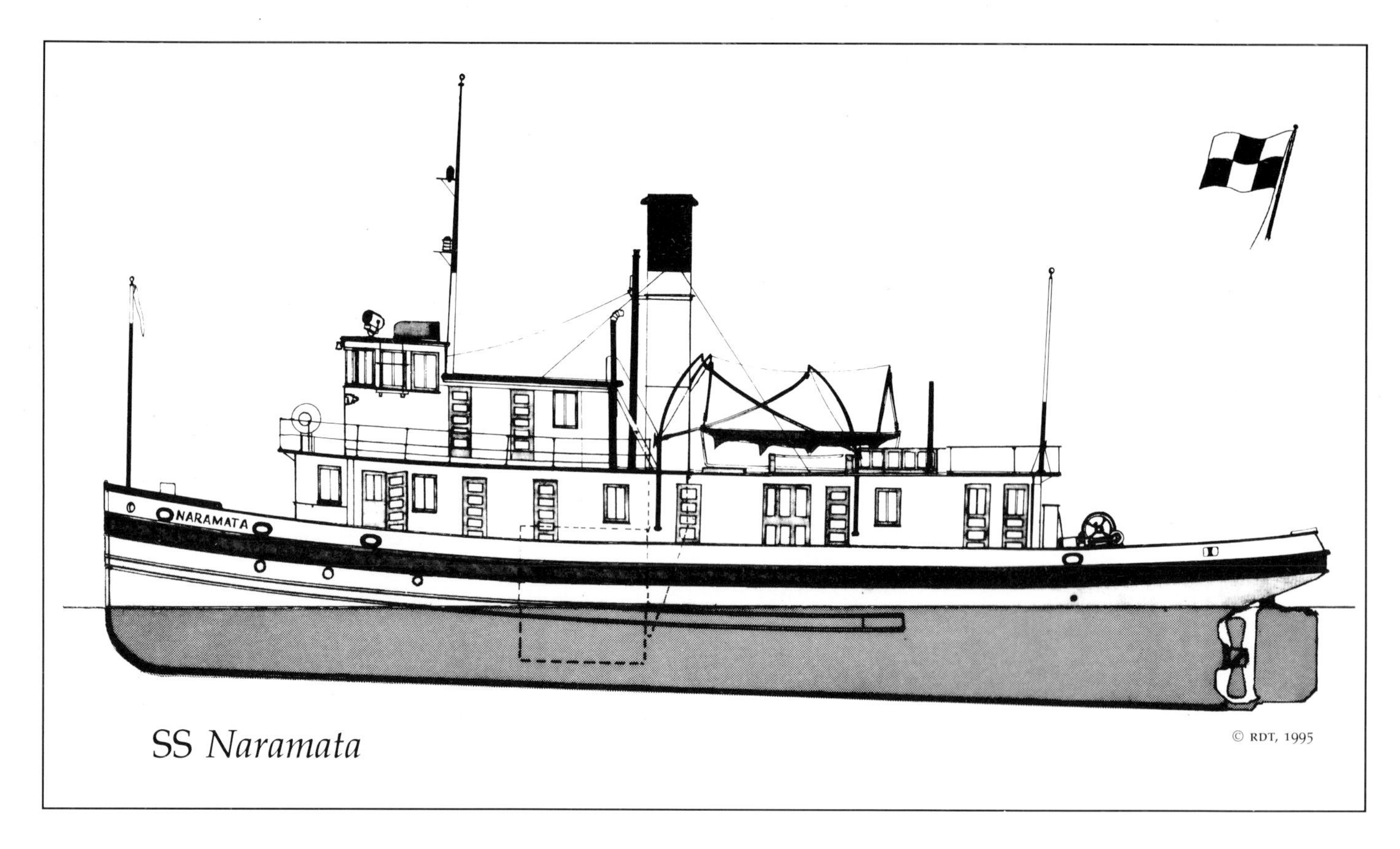

SS *Naramata*

© RDT, 1995

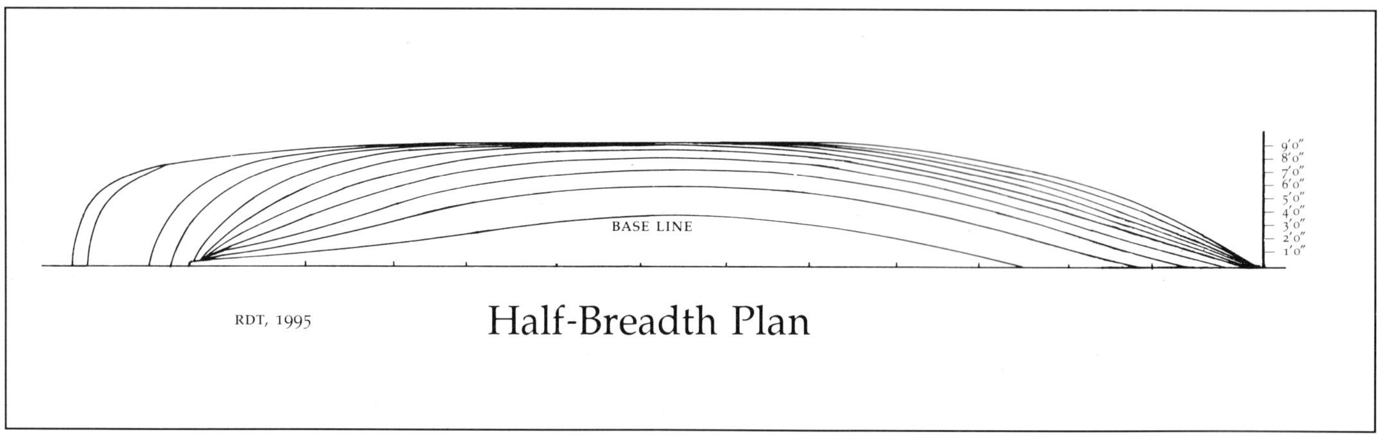

RDT, 1995

Half-Breadth Plan

Details of the Naramata

HULL DIMENSIONS & TONNAGE

Length: 89.8 feet* (98 feet overall; from specifications, but CPR 1938 records show 99.1 × 21.0 × 9.6 feet)
Breadth: 19.5 feet*
Depth: 8.0 feet* (Builder's plan shows 9.0 feet moulded depth)
Registered Tonnage: 73.67
Gross Tonnage: 149.94

* Registered Dimensions

Steel hull and deckhouse

REGISTRY / DATES

Official Number: 134271 (some records show 134217)
Port of Registry: Victoria, B.C.
Launch Date: April 20, 1914
Entered Service: 1914
Last Operated: August 1967
Sold by Canadian Pacific Railway to David Keffer, James Keffer, William Blackstock and Edward Walton, 1969; to Okanagan Landing Association; to Kettle Valley Railway Heritage Society, City of Penticton, 1991
Moved to Penticton: October 1, 1991

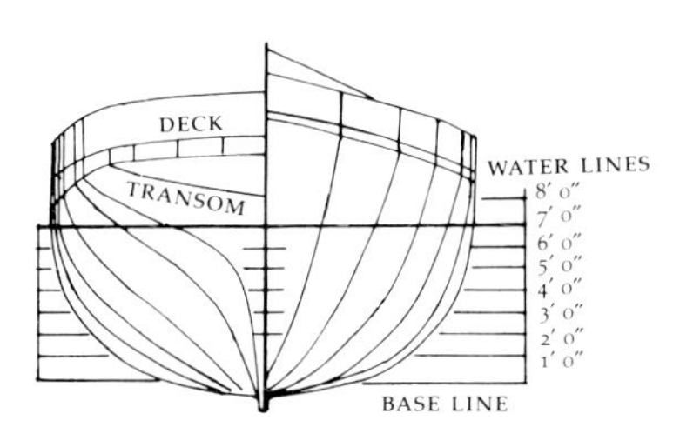

BUILDER (Hull, Engines and Boiler)

Prefabricated by Western Dry Dock & Ship Building Co., Port Arthur, Ontario, 1913, and assembled at Canadian Pacific Railway shipyard, Okanagan Landing, B.C.
Hull Number: 11
Cost: $43,000

ENGINES

Type: Fore-and-aft, Compound Jet-condensing, driving single screw
Dimensions: 12 & 26 × 18 inches
Propeller: Single, 6 feet 5 inches diameter, four blades
Horsepower: 27.3 nominal horsepower; 150 indicated horsepower
Number of Rudders: 1

BOILER

Type: Scotch Marine
Length: 10 feet 6 inches
Diameter: 9 feet 6 inches
Heating Surface: 1140 square feet
Grate Area: 33 square feet
Furnaces: 2, 36-inch diameter
Tubes: 110 – 3-inch × 8-foot plain tubes; 46 – 3-inch × 8-foot stay tubes
Working pressure: 160 pounds per square inch (1103 kPa)
Fuel: Coal
Speed: 12 miles per hour (19 km/h)
7 miles per hour (11.25 km/h) typical towing speed

PASSENGER CAPACITY & CREW

20 (1914)
22 (1916)
0 (by 1930s)

Crew: 11-13

CAPTAINS & CHIEF ENGINEERS

Captains of the *Naramata* during her long career (not shown in order of service) included: Matthew Pearce Reid, Joseph B. Weeks, Frank William Broughton, Donald MacFarlane, Malcolm McLeod, Jock M. McLeod, Otto Estabrooks, A. McKinnon, Robert S. Manning, John Thomson, Jack Thomson, Walter Spiller, Sam Podmoroff, Norman Nordstrom, Fred Barlow and Ed Schimpf.

Chief Engineers known or likely to have worked on *Naramata* include: John P. (Jack) Sutherland, W. Jacobs, T. W. Bracewell, W. Edwards, John F. McRae, John A. Williams, Johnny Miller, Charlie Verey, Samuel C. Bright, Robert Averal, C. S. Hingley, Herbert Gray, Ernest J. Hall, Jan Tambre, Martin Turunen, Robert Ingles, Harold Foote, Fred McKie and Elmer Hautaluoma.

SOURCES: Original Specifications, Steamship Registry and steamship inspection papers. Note that information varies slightly in some sources. Dimensions are in the form and units reported.

For conversion to metric measure:
1 foot (12 inches) = 30 cm; 1 inch = 2.54 cm

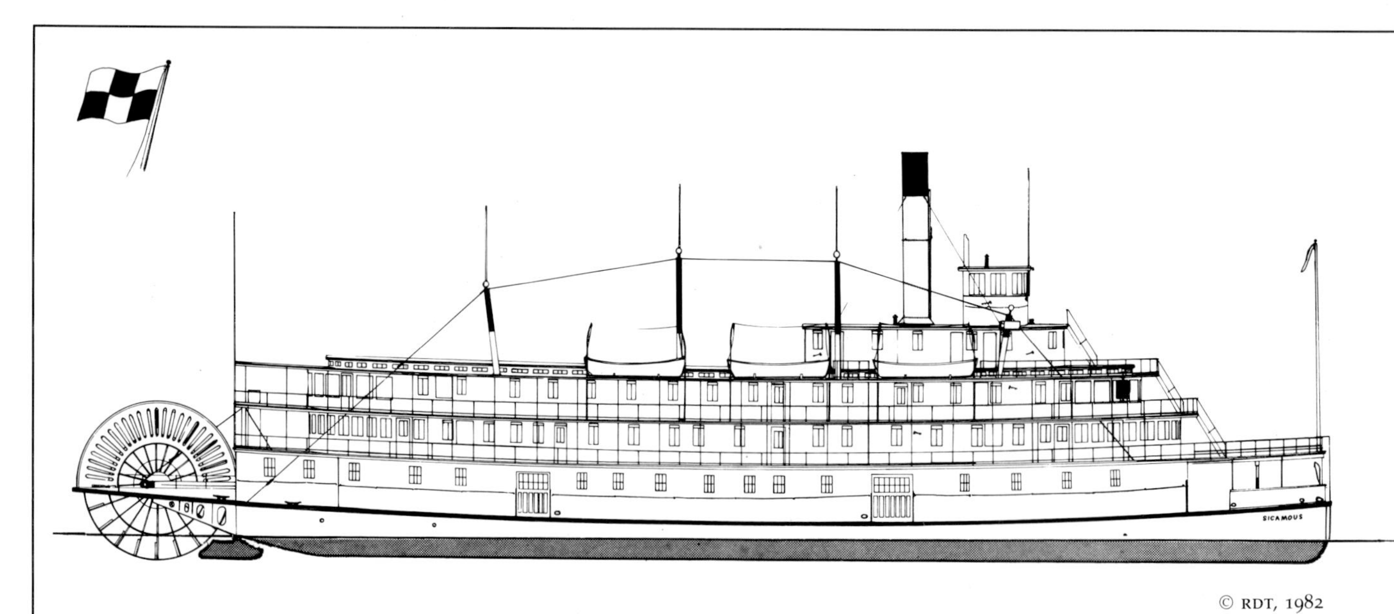

SS *Sicamous*, 1914

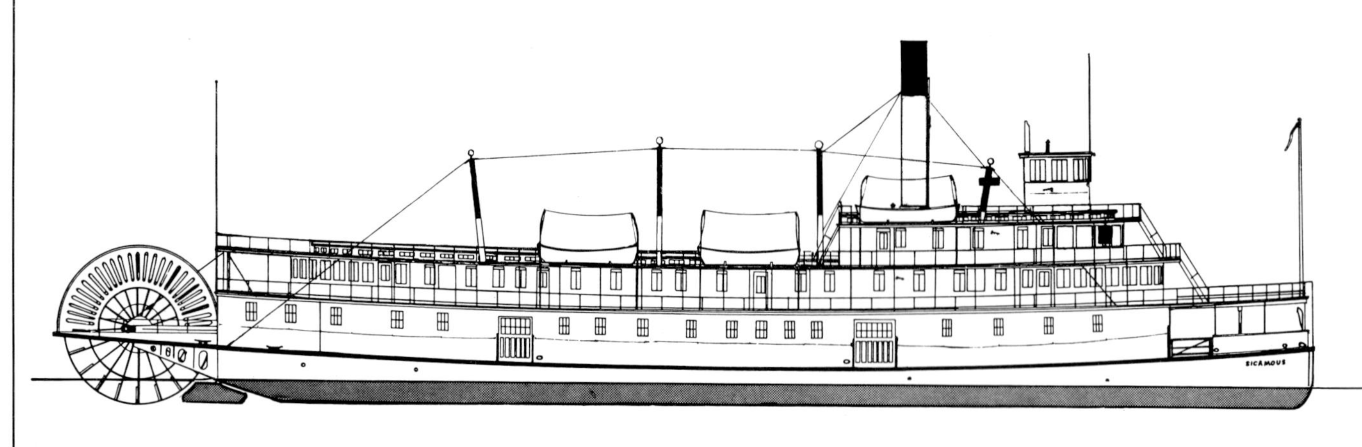

SS *Sicamous*, After 1935 Reconstruction

Details of the Sicamous

PASSENGER CAPACITY, ACCOMMODATIONS & CREW

500 (1914-16)
260 (1935)

Passenger capacity varied over the years. For example May 1916 records show 360 as the licence while the 1920 passenger tariff shows 400.

Dining Saloon Seating: 40-60 (five or six tables each with eight or ten chairs; early reports gave 70 as the capacity but this would have been crowded; 50 was typical).

Overnight Cabins: Most staterooms provided a single upper and double lower berth. Newspaper reports in 1914 show 40 staterooms but it is likely that some of these were finished for crew and other uses before the steamer was placed in service. Floor plans from 1920 show 36. There were four double suites (cabins 7, 8, 26 and 29) each with two upper single and two lower double berths, and two staterooms (Nos. 35 and 37) either of which could be opened en suite to a private bathroom.

36 (1920s)
10 (1935, after reconstruction)

TOTAL BERTHS: 80 (1920s)
20 (1935)

CREW: 31-33 (1914-1935) plus one mail and two express clerks
12 (1935-1937)

VEHICLE CAPACITY: 17 (1920s)
freight deck clearance of 7 feet

HULL DIMENSIONS & TONNAGE

Length: 200.5 feet* (registered);
228 feet (overall)
Breadth: 40.0 feet*;
Beam over guards: 43.0 feet
Depth: 8.0 feet*;
Average draught: 5.0 feet
Registered Tonnage: 994.38 tons
Gross Tonnage: 1786.65 tons

*Registered dimensions as reported in Steamship Registry. CPR fleet lists showed 202.5 × 39.1 × 7.5 feet.

Steel hull with wooden cabins.

REGISTRY / DATES

Official Number: 134276
Port of Registry: Victoria, B.C.†
Launch Date: May 19, 1914
Entered Service: June 12, 1914 (excursion); regular service, July 12, 1914
Last Passenger Service: July 1937
Sold to City of Penticton for $1.00, June 24, 1949
Moved to Penticton: August 27, 1951
Registry Closed: 1951

†Launching photos show "Sicamous of Vancouver" on the paddlebox, but the vessel was registered in Victoria and the paddlebox decoration was later changed. Port of Victoria registry documents from 1914 show Vancouver crossed out and Victoria inserted.

BUILDER (Hull, Engines and Boiler)

Hull prefabricated of steel by Western Dry Dock & Ship Building Co., Port Arthur, Ontario, 1913 (Hull No. 4) and assembled at Okanagan Landing where CPR crews built the cabins.

ENGINES

Type: Compound, Jet-condensing
Dimensions: 2–16⅜ and 2–35⅜ × 96-inch; piston valves on high-pressure cylinders, balanced slide valves on low-pressure cylinders.
Horsepower: 101.3 nominal horsepower;
1,000 indicated horsepower

BOILER

Type: Locomotive
Length: 23 feet 1 inch
Length including smokebox: 28 ft. 4 in.
Diameter: 90 inches
Length of tubes: 12 feet
Diameter of tubes: 2½ inches
Number of tubes: 320
Working pressure: 200 pounds per square inch (1379 kPa)
Firebox: 10 feet long × 7 feet 9 inches wide
Grate area: 70 square feet
Heating surface: 2750 square feet
Fire doors: 3
Steel: Carnegie Flange Steel
Smokestack: 60 feet long in two sections; 52-inch diameter
Fuel: Coal; designed to burn 3,800 lbs (1720 kg) of coal an hour under forced draft.

MAXIMUM SPEED

18-20 mph (29-32 km/h) [sources vary on the top speed and 20 knots was quoted in Penticton's *Herald* for the trial trip of June 12, 1914]
18 knots (32 km/h) [Certificate of Survey]

STERNWHEEL DIMENSIONS

Length of wheel bracket: 18 feet 9 inches
Diameter of wheel: 24 feet
Diameter of Shaft: 14 inches; made of nickel steel and weighing over 10 tons (10.2 tonnes), it was hollow through the centre with the inside diameter being 6 inches.
Length of paddle buckets: 20 feet
Number of paddle buckets: 20

NUMBER OF RUDDERS

3 (all forward of paddlewheel)

SOURCES: Original Specifications, Steamship Registry and steamship inspection papers. Note that information in some sources varies slightly. Dimensions are in the form and units reported.

For conversion to metric measure:
1 foot (12 inches) = 30 cm; 1 inch = 2.54 cm

suitable for families and two rooms, either of which could be opened onto a private bath that was located between them. All cabins were steam heated, had electric lights and were equipped with bells to call a steward. Running water to the staterooms was not provided. Lavatories were located on each deck and baths, one forward near the smoking room for men and one aft next to the ladies' saloon, were available, at a modest charge, on the saloon deck. Beautiful staircases, one just ahead of the dining saloon and the smokestack enclosure, and the other just aft of the saloon, connected the two decks and opened onto galleries adjoining the upper deck lounges. This gave the gallery deck three matching galleries overlooking the saloon deck.

Above the main passenger decks was the boat deck or Texas deck, which provided a large cabin for the captain and rooms for the mate, pilot, second engineer, watchman, cooks and waiters. Above this deck was the pilothouse. The steamer was reported to have cost $180,000, including $14,000 for furnishings, although CPR records place the value at $160,000.

While finishing work and the installation of the lifeboats remained to be done, the *Sicamous* was introduced to the people of the Okanagan in an inaugural excursion on June 12, 1914. A free train was provided for Vernon residents to travel to the Landing and from there they boarded the *Sicamous*, which was gaily decorated with flags. Many stops were made along the way to Penticton for people to board and inspect the steamer. "Captain Gore," recounted *The Kelowna Courier*, "repeating the procedure when the *Okanagan* made her maiden voyage, . . . shouted an invitation to all and sundry to board the boat and take a free trip. There was a hearty response, and young and old, male and female, to the estimated number of over 200 went aboard [at Kelowna] and after exploring the various decks settled themselves down to enjoy the voyage. . . . More were picked up at Peachland, making a total of about 400." Penticton's *Herald* gave the crowd as 500. If Okanagan residents thought the *Sicamous* looked tall, they were right. From the main deck to the top of the pilothouse was an impressive 53 feet (16.2 m) and the smokestack towered even higher.

After the successful trial trip down Okanagan Lake, the *Sicamous* returned to the shipyard for further finishing work. On July 1, 1914, she carried excursion passengers to Penticton for the Dominion Day celebrations but did not enter regular service until all of the lifeboats were finally installed. On July 12, 1914, the *Sicamous*, reported Vernon's *News*, "caused a great deal of excitement [at Naramata] when she made her first regular trip down the lake."

"They [*Bonnington*, *Nasookin* and *Sicamous*] had jet condensers so you had no engine exhaust noise . . . and only about 19 revs a minute for the paddlewheel. You could feel the thrust when the two cranks were up and coming down. You could feel that extra surge with the extra weight, once a revolution. You could actually count the revolutions by standing anywhere on the ship, you could just feel this little surge."
—CHIEF ENGINEER JOHN WILLIAMS, 1979

In full profile, the *Sicamous* reverses engines and pulls out into the lake after leaving Kelowna.—KELOWNA MUSEUM

A large crowd gathers at the Kelowna dock to meet the *Sicamous* on this warm, sunny day in 1915.—G. H. E. HUDSON, HERITAGE PHOTO CO-OP

At 24 feet in diameter, the paddlewheel of the *Sicamous*, turning from 15 to 20 revolutions a minute, made an impressive sight. —KELOWNA MUSEUM

Captain George L. Estabrooks was the *Sicamous*'s first regular captain. —*The Okanagan Semi-Weekly*

The officers and crew of the *Sicamous* posed for Penticton photographer Lumb Stocks in 1925. Captain Joe Weeks is standing in the middle of the top row in front of the pilothouse. Some of the crew members are wearing monogrammed sweaters while others have put on a shirt and tie for the occasion.—LUMB STOCKS, HERITAGE PHOTO CO-OP

Crew of the Sicamous 1918-1919

Captain
Mate
Deckhands (9)
Chief Engineer
Second Engineer
Firemen (3)
Coal passers (2)
Purser
Freight Clerk
Chief Steward
Stewards and cooks (10)
Total: 31

Steamer Days on the Lake

Although Okanagan Lake often may appear tranquil and peaceful, with few hazards for steamer navigation, it did present its challenges. The steamer route from Okanagan Landing to Penticton was nearly 65 miles (105 km) long with few sheltered bays or retreats in times of storms. The lake itself formed the bottom of a long, deep and narrow valley, and while up to 794 feet (242 metres) deep, was seldom over three miles (4.5 km) wide. Sudden storms and high winds could quickly turn the lake into a frothing mass of whitecaps, catch the steamers broadside and make steering extremely difficult. Winter ice could bring virtually all navigation to a halt, particularly at the southern end of the lake. Bone-numbing winter winds could cover the steamers with spray, which quickly froze into an encrusting layer of ice.

The *Sicamous* and the other Okanagan Lake steamers had many encounters with ice on the lake but none equalled the winter of 1915-16. By mid-January, temperatures were well below zero Fahrenheit (–17 degrees C), and strong winds were reported throughout the valley. Ice on the lake caused increasing problems for navigation and on the 14th, *The Summerland Review* reported that "it was necessary to cut the big boat, particularly the paddle wheel, out of the ice, and before she was free of the field of floating ice she was stalled several times." Winds from the north built up increasingly heavy ice off Penticton and by January 21, neither the *Sicamous* nor the tugs were able to reach Penticton and ended their runs at Summerland. One tug, probably the *Naramata*, had damaged its propeller in the ice while trying to make it to Penticton and had to return to Okanagan Landing for repairs. Ice also accumulated off Kelowna and by the end of January, the community of Naramata also was blocked. "That night [January 31]," recounted Summerland's *Review*, "the boat broke the ice, but did not attempt to return. Tuesday night the Summerland wharf was reached and the get-away made on Wednesday morning with considerable difficulty. But Captain Robertson was not going to be easily defeated, and that night again headed his ship for Summerland, but when within a short distance of his destination had to give up. For a time it looked as though the boat would be frozen in, but Thursday noon she got free and headed north again." Mail was taken by road to Crescent Beach and delivered over the ice by sleigh to the *Sicamous.* The *Naramata* and *Castlegar* worked with icebreaker barges to keep the

"I say the SS *Sicamous* provided a stateliness, a dignity, a peacefulness, a nobility, to our lives that many of us will never forget."—MARY ORR

"I remember the courtesy of the crew. She would be pulled into the shore where there was no dock or wharf in order to deliver one letter."
—JOE HARRIS, QUOTED IN *The Penticton Herald*, MAY 29, 1952, ON THE OPENING OF THE *Sicamous* BY THE PENTICTON GYROS

"They would stack the apple boxes up four high and then there was a stevedore, he had a hook, and he would tilt the boxes over and then you would run the lip of your hand truck under the boxes and we'd wheel it into the boat and stack them away. We did that all up and down the lake. Then we had to unload it all at Okanagan Landing."
—AUSTIN WILLET, DECKHAND, 1993

channels open but it was a constant battle. Conditions were not much better on the Kettle Valley Railway. Trains were encountering heavy snow in the mountains and were running as much as 26 hours late. Slides added to long delays for the trains through mid-February.

On the night of February 2, the *Sicamous* reached Peachland and Greata but stuck fast in the ice three miles from Summerland. "She could not go ahead and she could not back up to return to Peachland," reported Penticton's *Herald*. "Men were kept busy all night repairing the big paddle wheel. The mails and passengers were taken over the ice to Summerland." The lake was frozen for 13 miles (21 km) north of Penticton. The ice conditions continued to get worse and were so bad that the *Sicamous* could only be steamed as far south as Kelowna for over a week in mid-February but was later able to open a channel through to Peachland. Late that month conditions moderated and the ice began to break up. But it was not until March 11 that Captain Joe Weeks and his crew, with the *Naramata* pushing an icebreaker barge, forced a channel through to Penticton. Later that day Captain George Robertson brought the *Sicamous* into Penticton. After eight gruelling weeks, the service was reopened the length of the lake.

John Williams recalled another bad winter in the south Okanagan when he was second engineer of the *Sicamous*: "On Sunday morning we backed in from Summerland to Penticton using the sternwheel to break the ice, to break ourselves a path back into Penticton again, and I was horrified because I was the one who had the job of repairing the darn wheel when they got in there. To fix it took four or five hours. . . . I was horrified at them smashing up the wheel and the darn buckets instead of using the steel hull but they didn't want to take any chances with the hull. The wheel was expendable. That was our job and anything on the wheel, well, that was our [the engineers'] responsibility. It wasn't nice working out there six or seven feet above water with nothing to stand on except a little bit of a guardrail around the back or standing on the buckets."

High winds and waves were another hazard for the steamers. On November 11, 1924, the *Sicamous*, caught by a strong cross wind while approaching the Penticton wharf, was swung around and driven broadside onto the beach. "Waves were breaking against the side of the boat, throwing spray half way up the sides of the cabins on the second deck," reported Penticton's *Herald*. The crew finally rigged a cable from a piling on the wharf to the stern of the steamer and then ran it the length of the vessel to the capstan at the

Winter of 1916

"Constant repairs to the paddle wheel of the *Sicamous* are necessary, and . . . both tugs are continually experiencing propeller trouble. All are busy on the Kelowna channel."—*The Summerland Review*, FEBRUARY 18, 1916

"A heavy snowstorm made it impossible to perceive the boat's approach but . . . a nearby crushing and rumbling noise told that the long awaited had at last arrived. Just then the snow clouds lifted and not more than half a mile distant was the *Sicamous*, her prow thrust bravely into the ice, her stalwart form shaking perceptibly with the strain of combat, her funnel pouring forth a volume of smoke worthy of a man-of-war, and her paddle wheel kicking up a spray which would shame a good sized waterfall. . . . Upon close scrutiny one could see the ice being systematically broken down by the weight of the boat and being constantly pushed off to either side. . . .

"Soon everyone gained confidence and all walked out to get a view of the 'good ship *Sicamous*' at close range. She certainly presented an interesting sight—the paint long since disappeared from her lower hull, her sides were coated with several inches of ice and her paddle wheel looked like a miniature ice-castle. The upturned ice around her sides showed a thickness of five inches. . . ." —*The Penticton Herald*, FEBRUARY 24, 1916

"Our best thanks are due to the CPR tug *Naramata* and the mild chinooky weather of last week for opening up the lake, making it look a little more like a lake than a skating rink. Summerland is now enjoying the blessing of a daily boat service, and the tug has already plowed a way across to Naramata and made considerable headway toward Penticton."
—*The Penticton Herald*, MARCH 9, 1916

The *Naramata* steaming through heavy ice off Kelowna later in her career.
—BILL KNOWLES

To the engineers fell the task of maintaining the paddlewheel and keeping it clear of ice. A brief stop for the *Sicamous* provided an opportunity for chopping ice from around the wheel housing.—BILL TRIGGS COLLECTION

The *Okanagan* in the ice off Kelowna on February 25, 1922.
—MCEWAN PHOTO, KELOWNA MUSEUM

Keeping the steamer channel open was a job that fell to the tugs. Here the *Kelowna*, with an ice-breaking barge, steams through open water among the floes. The icebreakers had steel sheeting on their bows to protect them from damage.
—VERNON MUSEUM & ARCHIVES

At times it seemed the winters would never end. After weeks of battling ice, the steamer crews could be near exhaustion. Residents inspect the *Sicamous* from the ice at Kelowna.—PENTICTON MUSEUM

All Returned Men

— of —

Summerland, Peachland & Naramata

are requested to meet at the

C.P.R. Wharf, Summerland

at 10 o'clock; Tuesday Next,

to take part in Reception of The Prince of Wales. Wearing of Uniform optional.

S. A. MACDONALD, Sec'ty G.W.V.A.

—*The Summerland Review*, SEPTEMBER 26, 1919

"As the Prince stepped on to the deck of the *Sicamous* the children, joined by the crowd, sang the National Anthem, and again when he had reached the upper deck and the boat was pulling out the veterans cheered the Prince and sang 'For He's a Jolly Good Fellow.'"
—*The Summerland Review*, OCTOBER 3, 1919

"I remember very well how the skipper would blow the hooter without stopping half an hour before she was due to dock whenever he had wounded aboard. The men were always assured of a large crowd of people at the dockside to welcome them."—JOE HARRIS, QUOTED IN *The Penticton Herald*, MAY 29, 1952, ON THE OPENING OF THE *Sicamous* BY THE PENTICTON GYROS

Captains and Chief Engineers of the Sicamous

Captains included: John C. Gore, superintendent of the British Columbia Lake & River Service, who took charge of the steamer on her inaugural excursion, and George Ludlow Estabrooks, her first regular captain, in 1914; followed by William Kirby, 1915; and George Robertson from November 1915-1919 and again in 1921-1922 with J. A. McDonald being in command for 1920. Otto L. Estabrooks served as first officer and pilot under his father in 1914 and apparently was master briefly until the arrival of Captain Kirby from the Kootenays. Joseph Burrow Weeks had the longest period as captain, from 1923 to 1937. Most probably other captains worked or relieved for varying periods during the steamer's years of service. Chief Engineers on the *Sicamous* included William Jacobs, D. Stephens, D. H. Biggam, John F. McRae and P. H. Pearse. Chief Engineer Stephens is probably Dave Stephens, the BCL&RS's superintendent engineer, who normally oversaw operations from Nelson. A severe shortage of skilled engineers during the First World War likely led to his service on the *Sicamous*, filling in for younger men in the Lake & River Service who had joined the armed forces.

bow. "About 10 o'clock the wind and waves abated slightly and better progress with drawing the boat in was made, but it was not until after midnight that the *Sicamous* was berthed and the passengers and mail landed." Fortunately, no one was hurt and no damage done.

A more serious accident occurred in 1930 when early in January high winds and freezing weather hit the Okanagan. The *Sicamous* was caught several times in a cross wind and turned sideways as Captain Weeks tried to manoeuvre the steamer to the landings. In several places the crew had to warp the big sternwheeler into the dock using ropes and the steamer's winch. At Naramata, the wind caught her at the dock and she crashed into the wharf, breaking off dolphins and smashing in the side of the steamer. The mail room, express room and several staterooms were damaged. Fortunately, no one was hurt and the Okanagan Landing shipyard crew soon had repairs in hand. Later, the *Sicamous* was involved in a minor brush with one of the ferries that operated between Kelowna and Westbank.

Excursions were a part of sternwheeler operations on Okanagan Lake from the beginning of the *Aberdeen*'s career to the final voyage of the *Sicamous* under the auspices of the Penticton Gyro Club 44 years later. They were always popular. Dances, moonlight cruises on a hot summer night, church, Sunday school or fraternal organization picnics, the Kelowna Regatta, and service club special events all prompted excursions. They were a chance for relaxation, a break in routine, an opportunity to meet old friends and make new ones. Many a romance began on a dance cruise or picnic excursion.

Royalty, Governors-General and other well-known or highly-placed people also travelled on the *Sicamous* and the other sternwheelers. The most celebrated visit was undoubtedly the September 1919 tour of the Prince of Wales. After a reception in Penticton, the Royal party boarded the *Sicamous* and proceeded to Summerland where a brief stop was made. Several hundred people met the steamer at the "tastefully decorated" dock. Residents of Naramata and Peachland had come by boat for the short glimpse of the Prince. From there, the *Sicamous* steamed directly to Okanagan Landing where the party drove to Vernon for a round of events. After driving to Kelowna and more festivities, the Prince and his entourage reboarded the *Sicamous* and returned to Penticton that evening where they departed by train for Nelson.

Schedules for the *Sicamous* varied over the years, but typically her day began at Penticton with a departure, sometimes as early as 5:30, on a daily, except Sunday, service to Okanagan Landing and return. In some years, for the peak summer traffic, a Sunday boat also was operated. Passengers, mail and cargo were picked up at all of the important communities before the vessel pulled into Kelowna in mid-morning. Calls along the less-populous northern half of Okanagan Lake past Okanagan Centre were made at "flag stops" usually three times a week. In quiet times, the *Sicamous* would arrive at Okanagan Landing shortly after noon, meet the Vernon train, take on passengers, mail, cargo, supplies and coal before leaving southbound about an hour later. Stops were made again at the larger communities before she arrived back at Penticton in late afternoon. Passengers could spend the night on board at Penticton to avoid having to arrive at the docks before 5:00 a.m. In the summer months and for the duration of the fruit season the *Okanagan* worked on a complementary schedule to the smaller communities and packing houses. Operations of the *Okanagan* varied during the fruit season and in some years she was based overnight at Naramata or Gellatly. From there the steamer stopped at communities en route to and from Okanagan Landing.

Service standards on the *Sicamous* were never allowed to slip. Travellers always had the option of fine meals in the dining saloon and what was so often called "that wonderful CPR service." After lunch, passengers could enjoy tea, for 35 cents, served on the outer decks as they watched the scenery pass by. Before travel by car took away most local traffic, afternoon tea was particularly enjoyable for residents of communities such as Peachland returning from shopping trips to Kelowna. Weary travellers could relax in a hot bath for 50 cents before retiring to a steam-heated stateroom that, albeit modest, provided a warm bed with clean sheets and some thin-walled privacy. A traveller could rent a berth in a stateroom and share the room with another traveller or could rent the entire room at extra cost. Staterooms could also be rented for day trips if desired.

Unfortunately, few logs or records of the day-to-day operations of the *Sicamous* appear to have survived. An exception is a log page from a spring day in 1926 which Jack Petley was able to copy. The stops on the run as shown in the accompanying sidebar provide a fascinating insight into the steamer's routine. If this was a typical day on the *Sicamous* it is small wonder that within a few years the CPR was contemplating abandoning the passenger service. On this trip, passengers barely outnumbered crew and freight totalled less than one carload. Fortunately mail and express revenue would

Captain Joe Weeks, remembered with great affection by travellers, officers, crew and residents of the Okanagan, was master of the *Sicamous* from 1923 until her last voyage in 1937. He welcomed passengers to the pilothouse where he kept a log or guest book for visitors.
—VERNON MUSEUM & ARCHIVES

Master's Log, April 13, 1926

Penticton	6.20	
Naramata	6.58	7.03
Summerland	7.14	7.20
Peachland	8.20	8.25
Westbank	8.55	9.00
Kelowna	9.36	9.49
O.K. Centre	10.44	10.47
O.K. Landing	11.55	13.35
Killiney	14.09	14.15
Ewing	14.49	14.53
Fintry	14.33	14.40
O.K. Centre	15.10	15.14
Kelowna	16.11	16.30
Westbank	17.10	17.14
Peachland	17.45	17.50
Summerland	18.46	19.00
Naramata	19.10	19.15
Penticton	19.47	

Captain Weeks noted the following in the log:

Stage of Water at O.K. Landing: 100.20.
Weather: Calm, clear and beautiful.
Fuel taken at Okanagan Ldg.: Tons 14½.
Tons of cargo: Up trip, 4; Down trip, 15.
Number of passengers: Up trip, 30; Down trip, 34.
Number of crew all told 31.

—ARROW LAKES HISTORICAL SOCIETY

have helped balance the accounts. For the crew on this April day, there was little time to rest with six stops on the northbound trip between Penticton and Okanagan Landing and nine on the return. Sometimes stopping for less than five minutes, the *Sicamous* pulled in only long enough to exchange mail, express and freight and pick up or drop off a few passengers.

Planned in the expansive years before the First World War but frequently delayed, a Canadian National Railways branch line was finally completed to Kelowna from Kamloops in September 1925. The route made use of new CNR trackage and parts of the CPR into the valley. The Canadian Pacific gained running rights into Kelowna for freight trains and service began on August 1, 1926. Increasingly, fruit shipments were moved through Kelowna rather than by the slower steamer or tug and barge route through Okanagan Landing. After 1926, the barge service operated from Kelowna but the CPR passenger steamers still connected with passenger trains at the Landing. Canadian National also developed a passenger service, using the motor vessel *Pentowna*, built in 1926, and a tug and barge operation on Okanagan Lake. CN built transfer slips at Naramata, which the CPR also used, and at Peachland, Westbank, Penticton and Kelowna. The move of the main barge terminal to Kelowna reduced the length of the trips and made it feasible to operate the service without the *Castlegar*, which was retired in 1925. The barge service was left to the *Naramata* and to the *Kelowna* which had been added to the fleet in 1920. Meanwhile, the *Aberdeen* and *Kaleden* had been retired.

Highway construction, the improvement of ferry service between Kelowna and Westbank, and increasing bus and automobile competition during the 1920s hit the steamer service a heavy blow. The onset of the Depression of the 1930s only made matters worse. Traffic fell off dramatically and the CPR considered ending passenger service on Okanagan Lake as losses mounted. The company went so far as to announce the termination of service, as the *Sicamous* reached the end of its four-year refit cycle, in the spring of 1931. On many days, there were very few passengers on board to enjoy the spacious lounges and cabins on the *Sicamous*. However, after a meeting between General Superintendent Charles A. Cotterell, other CPR officials and representatives from communities, and the Vernon Board of Trade early in May 1931, the CPR agreed to continue the service. The small communities along the west shore of Okanagan Lake, north of Kelowna, had been particularly worried because roads there were still rudimentary or non-existent. Nonetheless, the future of the *Sicamous* was far from assured; the service had a reprieve and no one could imagine how bad the Depression was to become. Cuts were inevitable and the *Okanagan* was retired in 1934 and eventually scrapped with only some small sections of cabin, used as beach houses, remaining. Fortunately, the ladies' saloon survived to be acquired by the Sicamous Society in 1994.

Elsewhere, the sternwheelers were also falling to road and rail development. The CPR proposed abandonment of service on Upper Arrow Lake but in the end continued the operations of the *Minto*. However, the *Bonnington* was withdrawn in 1931 and the *Nasookin*, along with the *Kuskanook*, was replaced by expanded rail service and roads along Kootenay Lake. The *Nasookin*, first under lease and later purchased by the provincial government, was to continue in operation but as an automobile ferry across Kootenay Lake until retired in 1947. Of the three sisterships, only the *Sicamous* remained in active CPR service as the Depression of the 1930s worsened.

For 20 years after her completion, the *Sicamous* shared the busy summer and early fall season with the *Okanagan*. *Sicamous* handled the mail service and traffic from the larger centres while *Okanagan* called at the smaller packing houses and landings. With their speed and large cargo capacity, they moved thousands of crates of premium fruit. The *Okanagan* also filled in when *Sicamous* was being overhauled but it was expensive to maintain two vessels for these brief busy times. In this 1920s portrait, probably taken by Dorothy Gellatly, they pass off Peachland.—HERITAGE PHOTO CO-OP

"My mother and my Aunt Laura devised a little holiday by taking the boat as far as Kelowna and then we would go for a walk in the park and around the stores and then go to the original Royal Anne Hotel for lunch. We always had Royal Anne cherries and pickled olives on the stem. Then the *Sicamous* would go on up to the Landing and turn around and be back at the Kelowna wharf at about 4:00 o'clock in the afternoon. So then we would get on and have our dinner on board and be back to Penticton about 7:30 in the evening which made a very nice day's outing."
—BONNIE DAFOE, 1993

"To travel hopefully is better than to arrive . . ."—ROBERT LOUIS STEVENSON

Travel on the *Sicamous* was an experience that many remembered for a lifetime. For Gib Kennedy, above right, who would later work for Canadian Pacific and become a historian of its British Columbia operations, a trip on *Sicamous* as a three-year-old in 1919 was an adventure. For families and friends, far right, summer outings for picnics and shopping in Kelowna or Penticton, such as this one in July 1926, were the best of times. But for soldiers on board *Sicamous* and friends or family on the dock at Westbank, at right, during the First World War the excitement of travel was tempered by the reality of growing casualty lists in the newspapers and reports from the Western Front.
—W. GIBSON KENNEDY COLLECTION; VERNON MUSEUM & ARCHIVES; DOROTHY GELLATLY COLLECTION, HERITAGE PHOTO CO-OP

Dining Room Service

Canadian Pacific Railway steamship, dining car and hotel meals were renowned for their quality and high standards of service. Meals on the *Sicamous* often featured the fresh fruit and produce of the Okanagan. In 1915, all meals on the *Sicamous* and the *Okanagan* were 75 cents while on the *Aberdeen*, working mostly on the fruit harvest, meals were just 50 cents. Children under 12 years of age ate for 50 cents except on the *Aberdeen* where the cost was 25 cents. By the 1920s prices had risen to $1.00 for breakfast, $1.25 for lunch and $1.50 for dinner with children dining for half price. B.C. Lake & River Service menus offered a modest, but carefully prepared, choice of entreés, beverages and desserts. Standards were carefully maintained and normally one waiter would be assigned to each of the large tables, set with linen tablecloths and napkins and the CPR's special silver and monogrammed china in full, elegant array.

Ruth McGregor recalled a trip with school friends in the 1930s. "It was either provincial or interior basketball championships that we were going to. It was really fun and a good group of kids that were travelling that way. We knew that we were just being spoiled too, travelling on the *Sicamous.* Almost a private boat to take us up there. We caught the train at the other end and then went by pass to Revelstoke." Posed on the *Sicamous* are: Ruth, at the top; Estelle Tupper; Harriet Pentice; and Louise Nagle.—RUTH HANSEN MCGREGOR

Stewards wait for passengers to take their seats in *Sicamous*'s elegant dining saloon. —DOUG ELLIOTT PHOTO, PENTICTON MUSEUM

Mail on the Boats

Connecting at Okanagan Landing with the S&O Railway, the *Aberdeen* carried the mail to and from settlements along the lake, beginning in 1893. In 1894 mail cars, with railway mail clerks on board, started running over the S&O and the mail was transferred to the *Aberdeen*. Revenue from mail on the boats was important and the CPR received, for example, $844 for the tri-weekly service in 1895. Pursers likely handled mail in the early days and a "way mail" pouch for mail picked up along the route probably was carried on board. With the arrival of the *Okanagan*, service improved generally and in 1909, letters were being franked "Pen. & Ok. Ldg. R.P.O. Str. Okanagan," signifying that a railway post office was operated between Penticton and Okanagan Landing although details of the level of service are uncertain. Newspaper reports indicate that a clerk was appointed to the *Okanagan* early in 1912 and mail rooms were added to the steamers at about this time. Mail clerks sorted and handled mail on the steamers until 1935 when the *Sicamous* was taken off her regular passenger run. Railway post office service was established between Kelowna and Sicamous on the CPR and the days of mail by steamer in the Okanagan ended.

Okanagan Landing, above left, was a bustling, busy place when the train from Vernon met the *Sicamous*. By the early 1920s, private automobiles, very much in evidence in this picture, were taking many passengers away from the boats.
—G. H. E. HUDSON,
HERITAGE PHOTO CO-OP

Sicamous off Squally Point near Peachland.—HARRY SINCLAIR

Coaling at the Landing

"At Okanagan Landing, half the crew had to load coal for fuel. She took 17 tons of coal for a 24-hour trip. And that all had to be loaded within the hour that they stayed at Okanagan Landing. There were two men shovelling the coal; it was fine dusty coal, and the third man would wheel the barrow onto the boat and dump the coal down in front of where the boiler is. The other two or three of us would take on any freight."
—AUSTIN WILLET, DECKHAND, 1993

Cars of coal at dockside await the deckhands at Okanagan Landing in this 1924 photo. Standing on deck is a travelling psychic who called himself "Prince Chandu."—GEORGE MEERES, HERITAGE PHOTO CO-OP

"From the point of view of good business, it is generally admitted that the Company would be perfectly justified in withdrawing this service, which would be the easiest way out of the position in which it finds itself today. However, I cannot bring myself to the point of making such a recommendation, even though it may be argued that it would only be taking away a convenience from a comparatively small number of people."
—GENERAL SUPERINTENDENT C. A. COTTERELL, WRITING TO VERNON BOARD OF TRADE PRESIDENT G. O. NESBITT, MAY 1931

VICTORIA

The Sicamous Rebuilt and the Last Excursion

The Kelowna Board of Trade excursion of January 5, 1935, was an opportunity to pay tribute to Captain Weeks and his crew and mark the inevitable passing of an era of travel in the Okanagan Valley. Like most others in the Okanagan by the mid-1930s, few of these people would normally take the steamer if they could travel instead by automobile.—PENTICTON MUSEUM

As passenger traffic declined on the *Sicamous* during the Depression years from the already low levels of the 1920s, few people made use of the overnight accommodations and with the beginning of improved rail service to Kelowna it was clear that they were no longer needed. Substantial losses each month on the *Sicamous* showed no signs of being reversed. Moreover, the CPR had scheduled the beginning of its own passenger trains to Kelowna over the CNR; the end was at hand for the passenger steamer service. At the same time, the CPR steamers would no longer carry the mail with all service being handled either by train or over the highways. CPR Assistant General Manager from Vancouver, C. A. Cotterell, announced in December 1934 that the *Sicamous* was to be taken out of service on January 5, 1935, and given a major overhaul at Okanagan Landing. Her function would be to handle fruit and cargo and to carry day passengers during the summer months. Through-train service to Kelowna was to begin on Monday, January 7, so there would be no interruption for travellers or mail. As a freight boat it seemed possible that the *Sicamous* could pay her way.

The Kelowna Board of Trade hastily organized a party of 50 to 60 people representing many pioneers of the district to board the *Sicamous* on Saturday, January 5, and make the trip to the Landing and back as a farewell gesture and to present a tribute to Captain Weeks. Among those present were Board of Trade President David Chapman, Mayor W. R. Trench, CPR Revelstoke Superintendent J. J. Horn, and a young hardware merchant named W. A. C. Bennett who would one day become premier of the province. Captain Weeks was presented with a pen and pencil desk set. Toasts followed and the group sang "For He's A Jolly Good Fellow." Purser A. Watson was given a large package of cigars and cigarettes for the crew. When the steamer docked at Kelowna on the return trip, the passengers, reported Kelowna's *Courier*, "circled the rail in the cabin overlooking the saloon, joined hands and sang 'Auld Lang Syne' followed by cheers for Captain Weeks, who afterwards stationed himself at the gangplank and shook hands with all passengers as they disembarked."

While the days of the sternwheelers were drawing to a close, the tug and barge service was still essential. The *Kelowna* awaits as the crew unloads refrigerator cars from the barge at Peachland. —HERITAGE PHOTO CO-OP

The *Sicamous* and her sisterships had experienced some problems in cross winds because their tall decks of cabins acted somewhat like sails and made steering difficult. Plans had been developed to remove the upper cabins to improve handling and at the same time reduce maintenance costs. A similar reconstruction had been done to the *Nasookin* on Kootenay Lake after its sale to the provincial government for use as an automobile ferry. On January 10, 1935, the *Sicamous* was winched up onto the ways of the Okanagan Landing shipyard for the start of the reconstruction. There, the experienced crew began the task of removing some of the upper cabins. Men were brought in from the other major yards at Nelson and Nakusp and the alterations progressed rapidly. John Stobo, master builder from Nelson, was in charge and Arthur Weston was the Okanagan Landing shipyard foreman.

The job was complicated. Both the freight deck and the saloon deck were left essentially unaltered. However, the rear two-thirds of the upper deck was removed completely as was the short Texas deck under the pilothouse. The top open deck, with its clerestory windows, which could be described as the ceiling of the upper deck, was lowered onto the saloon deck. The removal of the upper decks reduced the overnight staterooms from 35 to just 12. This work also made a dramatic change in the dining room. The upper deck gallery, with the surrounding overnight cabins, was eliminated. Instead, the dining saloon was just one deck high, although still spacious and pleasant. During the refit the passenger accommodations were given a fresh coat of paint, the hull was scraped and painted, the engines were overhauled and parts of the paddlewheel were replaced. Whether at this time or earlier is unclear, but the interior decoration was simplified with white covering many of the highlights of gold and light blues that once brightened the cabins.

With the removal of the Texas deck cabins, the pilothouse was moved forward and lowered onto the top of the lounge that formed the front of the upper deck. Other features of the vessel had to be altered as well. The tall hogposts, which helped strengthen the hull, had to be lowered, the hog-chains modified, and the smokestack and the lifeboat davits shortened.

Re-launched on May 4, 1935, the *Sicamous* returned to service early on July 3. Her main function was to operate as a freight boat, as the *Okanagan* had once done, assisting with moving the fruit crop and carrying the few passengers who wished a leisurely trip on Okanagan Lake. Once again Captain Joe Weeks was in command and Otto Estabrooks was first officer. The reduced crew size of just 12 showed the effects of cutting out the dining saloon and overnight services. Based at Okanagan Landing, she ran as far south as Penticton on a daily, except Sunday, basis during the summer

Ed Wanstall, second from right, gave his camera to Stan Leno to take this picture of some of the crew as work on the *Sicamous* neared completion. The other men are Ed Robertson, Dave Thompson, an unidentified shipyard hand, and "Summerland Slim."
—ED WANSTALL COLLECTION

"They were going to lift the pilothouse and swing it around and lower it down to the ground. But as it lifted, the [sheerlegs] broke. They had to mend it all and then they swung it down. Then the carpenters just cut that top deck right off in pieces. Sheeted in what they cut off and put the pilothouse back on. Then that whole deck had to be recanvassed. . . . I enjoyed it over there at the Landing."
—ED WANSTALL, SHIPYARD WORKER, 1988

Some of the shipyard crew at Okanagan Landing during the reconstruction of the *Sicamous* in 1935.—ED WANSTALL

In 1991, Doug Griffith examines marks on the canvas decking showing the original position of the Texas deck before its removal in 1935.—ROBERT D. TURNER

months and through the fall fruit season. The days were long, beginning with an Okanagan Landing departure of 7:00 a.m. and not ending until the steamer returned at about 2:00 the next morning. Running north in the evening and at night meant that the fruit would be cooler on the unrefrigerated cargo deck of the *Sicamous*. During the winter and spring she was tied up at Okanagan Landing. In 1936, operations followed the same general pattern until the steamer was withdrawn on October 11. However, with the expanded rail service to Kelowna and the increasing use of automobiles, the CPR found that running the *Sicamous*, even as a freight boat, was not paying. Tugs and barges could handle the freight with more flexibility and at lower cost.

In July 1937 the Gyro Club of Kelowna was hosting a district convention and had guests from British Columbia, Washington and Oregon. As part of the convention, the Penticton Gyro Club chartered the *Sicamous* and sponsored what was to be the last passenger-carrying voyage for the steamer. The Gyro Club, whose symbol represents long-term stability, is an international organization promoting friendship and whose member clubs often undertake civic projects. For the Gyros, the *Sicamous* was an ideal venue for part of their convention. As was so often the case, Lumb Stocks was on hand to photograph the event and Captain Joe Weeks was in command. Penticton members and guests drove by private car to Kelowna to board the *Sicamous* along with the many others who were attending the meetings. After the leisurely run down the lake, the passengers enjoyed a banquet at Penticton's Incola Hotel. Later in the day, the *Sicamous* took the group back to Kelowna and soon steamed to Okanagan Landing where she was tied up, never to operate in passenger service again.

The company apparently planned on operating the *Sicamous* in freight service that summer but it appears that cool weather in August, which slowed the ripening of the fruit crop, and the use of the two tugs made running the big sternwheeler unnecessary. She was to remain at Okanagan Landing, largely forgotten, for the next 14 years. Fortunately, the tug and barge service was productive and profitable.

The Fruit Boat Sicamous

Sicamous returned to service, after reconstruction, in time for the summer fruit rush of 1935. Laid up over the winter, she again operated in 1936. The beautiful days of summer found her once again at Penticton in weather that even the Great Depression could not tarnish.
—LUMB STOCKS, HERITAGE PHOTO CO-OP

The Gyro Excursion of July 23, 1937

The last excursion on the *Sicamous* was sponsored by the Penticton Gyro Club in July 1937. It proved to be a highlight of the club's regional convention and was a fitting end for the steaming days of the sternwheeler.—KELOWNA MUSEUM

"We made an Ogopogo out of large truck tires and we had it hauled past the *Sicamous* when it was coming into Penticton. . . . They thoroughly enjoyed [the excursion] and when they saw Ogopogo, they all rushed over to the side of the boat but it didn't swamp us!"
—CLEMENT BATTYE, 1991,
WHO WAS GYRO PRESIDENT IN 1957

The Fruit Harvest and Tugboat Days of the Naramata

"In the fruit season [in the early 1920s] there were three tugs and six barges and the two sternwheelers, the *Sicamous* and the *Okanagan*. They were all running . . . from August until the end of October when they generally seemed to tie up . . . the three fruit boats."
—CHIEF ENGINEER JOHN WILLIAMS, 1979

Moving fruit from the many packing houses around the lake to the railway at Okanagan Landing was the main purpose of the tug and barge services, and also an important source of traffic for the sternwheelers on Okanagan Lake. Speed was important in handling the highly valuable and perishable fresh fruits, particularly in the heat of an Okanagan summer. Fruit was often shipped at night to avoid transit during the heat of the day.

When the CNR extended its trackage to Kelowna in 1925 and the CPR gained operating rights to the city, it became the preferred point for transferring the cars of fruit from the barges to the railway. Shipments could be expedited over the railway much faster than they could be handled on the lake by the tugs. Okanagan Landing faded in importance and in 1927 the barge service shifted its focus to Kelowna. By 1933, scheduled service to the Landing had dwindled to a mixed train from Sicamous and this ran only until that December. Passengers for the *Sicamous* continued to pass through the Landing until she was withdrawn for reconstruction in 1935 and the agency was closed. The few passengers who travelled on the *Sicamous* during her short career as a freight boat would have simply purchased their tickets from the purser on board. By 1940, no revenue was credited to the line and, without any public opposition, approval was granted to remove the tracks.

The once busy settlement became a quiet community with the shipyard the major centre of activity. When new barges were required the busy times returned. In 1936 a 15-car, three-track, steel transfer barge from Kootenay Lake was cut in sections and moved to Okanagan Landing where it was reassembled as a 10-car, two-track barge. New wooden barges were built in 1934, 1940, 1945 and a final steel barge was assembled in 1951.

Chief Engineer John Williams recalled the fruit seasons when he served as second engineer on the sternwheeler *Okanagan*, and on the tugs during the 1920s. "The *Naramata* was the steel tug that was the year-round tug. The *Kelowna* and the *Castlegar* were wooden tugs and were the ones that came out for the fruit season. The *Okanagan* used to run from Naramata to Okanagan Landing and we didn't go to Penticton. The *Sicamous* handled any small fruit coming out of Penticton. We started out about 2:30 in the

After the retirement of the *Sicamous* in 1937, *Naramata* still had 30 years of service ahead on Okanagan Lake.
—ERIC SISMEY

morning from Naramata and we'd get to the Landing around noon. In the height of the season we'd maybe have five, six, seven carloads of fruit. A carload was around 840 boxes of apples, that's 840, 40-pound [18-kg] boxes. The *Okanagan* would be full of apples by the time we got to the Landing in the height of the fruit season. We'd pick up a carload here, a carload there . . . Peachland, Westbank, Kelowna, Okanagan Centre, they were the regulars; there were a few other places . . . Fintry was one at the top end of the lake . . . we'd go in to pick up.

"It was all handled by hand, that is a man with a hand truck. They used to carry five or six boxes at a time, quite a chore to load a car of fruit; it took a little while. They'd have seven or eight men trucking and somebody counting. Of course every box had to be counted all the time. We would get in . . . roughly the same time as the *Sicamous* got in [at Okanagan Landing] on the passenger run and they would unload and then coal up and back we would go, back down the lake."

The tugs usually pushed the barges. Depending on traffic, weather conditions and the power of the tug, one or two barges were handled at a time. Normally, a tug was tied tightly to the side of a barge, towards the rear. When two barges were handled, they were tied together at the front and the tug was wedged in between them at the back forming a V pattern with the tug in position to push and manoeuvre the barges effectively.

The tugs seldom towed barges as was commonly done on the coast except in very rough weather. Towing eliminated the danger of the tug and barge rolling against each other in a storm. However, the short distances between transfer slips or landings, the normally calm water, and the confined nature of the lake made pushing the barges safer and much more efficient. When they came to one of the ferry slips, it was relatively simple to ease one barge into the slip, unload and load freight cars and then shift over the second barge into the slip. "They were lashed in there tight," noted John Williams. "They had winches on the front so that it was more or less one solid piece of equipment; it used to steer all right."

The earlier wooden barges were each capable of carrying eight freight cars; the steel barges, used in the later years of the service, could each carry 10 cars. A tug, with two fully loaded barges, could move the equivalent of a 16-20-car train.

After 22 years, during which no new vessels were ordered, a new tug was built in 1947. This was a diesel-powered, steel-hulled vessel named the

John Williams worked on nearly all the sternwheelers and tugs in the CPR's Lake & River Service, eventually becoming a chief engineer. He became second engineer in 1922 and worked on the *Okanagan*, *Castlegar*, *Kelowna*, *Naramata* and *Sicamous* and was chief on the *Kelowna* and *Naramata* in 1930 and finally *Moyie* in 1930-31 before transferring to Cominco at Trail.
—JOHN WILLIAMS COLLECTION

The *Naramata* and the other Canadian Pacific tugs were painted an attractive green, black, white and buff colour scheme. This photo shows the tug at Kelowna on August 18, 1958. After nearly 45 years of service on Okanagan Lake, the *Naramata* was still a hard-working vessel and in remarkably good condition. The small cabin at the back, above the main deckhouse, was added for extra crew accommodations.
—MAURICE CHANDLER

Okanagan. With its arrival, the *Naramata* became the relief vessel and, with the *Kelowna*, was used mostly during the fruit season. Winter weather sometimes brought her out to assist in keeping the barge shipments moving and she was also steamed up when the *Okanagan* was being given maintenance in the spring.

The work of the tugs was unglamorous but essential to the lives of countless fruit ranchers in the Okanagan. Over their long years of service, the tugs and barges moved thousands of carloads of fruit to the railway at Okanagan Landing and later to Kelowna. The rapid shipment of fruit and other produce from the Okanagan was critical to the agricultural industry and it also generated heavy traffic and earned substantial revenue for the CPR. During the peak of the fruit season, the CPR operated a special fast freight called the *Okanagan* between Kelowna and Winnipeg. In 1952, for example, it made 54 trips and connected closely at Kelowna with the arrival of the tug *Okanagan* at 2:00 each afternoon. The 10 major lake shipping points were contributing about 20,000 tons of freight with gross earnings averaging around $600,000 during the late 1940s and early 1950s; when costs and revenues were broken down further, the net earnings of the lake service were still substantial and reflected a level of profitability that would have made many railroads envious.

In later years, when the main service was between Kelowna, at the end of the railway, and Penticton, "we would take the empties down and bring back the loads which had started all the way from Osoyoos, Haynes, and Oliver," Captain Walter Spiller recalled. "They put them on the barge and we went through to Kelowna and we'd have to stop along the lake and pick up cars which were in readiness at either Naramata, Summerland or Westbank. . . . During the busy season the *Naramata* had a steel barge and a wooden barge that [in total] carried 18 cars. The busiest I would say was Summerland; Naramata was a close second. Those benches on each side of the lake were full of orchards."

The *Kelowna* was retired in 1956 and as traffic moved increasingly by road the *Naramata* was used less and less. In her last years she usually operated for about one month during April, May and June and again at the height of the fruit rush from mid-August sometimes as late as mid-October. The years of hard service took their toll on the old tug and a 1962 survey concluded that in five years a new engine, at a cost of about $100,000, would be needed to keep her in service. Moreover, major upgrading would be needed for the entire

The *Kelowna* was a familiar sight pushing barges on Okanagan Lake between 1920 and 1956.—ROBERT MANNING

"That was a lot of hard work; breaking a channel in the ice off Peachland. We just ran her into the ice. We had to work at it all day, hours and hours, breaking ice. It was noisy, like being inside of a drum. There was no sleep at all. The next morning we had to get out and go over the side and chip away. We finally got her, the *Naramata*, free and away we went."
—BILL GUTTRIDGE, FIRST OFFICER, 1994

lake service. The *Naramata*'s final trips were in August 1967 to assist with the fruit rush under Captain Norman Nordstrom. On August 29, 1967, she left Kelowna at 6:20 a.m., steamed north and arrived at Okanagan Landing at 9:00. The tug was tied up and the crew was released at noon that day, ending the *Naramata*'s steaming days for the Canadian Pacific. Two years later, the *Naramata* was offered for sale, "as is, where is." Sold to David Keffer, James Keffer, William Blackstock and Edward Walton, the tug was moved to Fintry.

The *Okanagan* continued working the transfer barges until May 31, 1972, when Captain Sam Podmoroff brought her into the Penticton transfer slip for the last time. Road development, changes in the fruit packing, processing and marketing businesses and the use of mechanical refrigerator cars all contributed to the demise of the service. Even in the early 1970s, it was not so much uneconomical as it was obsolete. The *Okanagan* was tied up at Penticton until November 3, when it was sold and moved to Fintry. Several attempts were made to use the big tug over the next 20 years but none proved successful. CNR lake service ended the next year; the tugs *Pentowna* and *CN No. 6* were both sold and eventually preserved. The days of railway tug and barge service on Okanagan Lake had quietly come to an end.

Norman Nordstrom and Sam Podmoroff were the senior captains in the last years of the tug and barge operation. Captain Nordstrom ran *Naramata* during its last summer season.
—NORMAN NORDSTROM COLLECTION

"The *Naramata* tied up about 1967. We took her to the Landing and tied her up and she never turned a wheel since."
—CAPTAIN NORMAN NORDSTROM, 1994

The diesel tug *Okanagan*, built in 1947, took over from the *Naramata* as the regular tug on the barge service leaving the ageing steam tug as the relief vessel and for use during the peak traffic periods of the fruit season. This dramatic photo shows *Okanagan* at Penticton in cold, winter conditions.—DAVE CLARK, JACK PETLEY COLLECTION

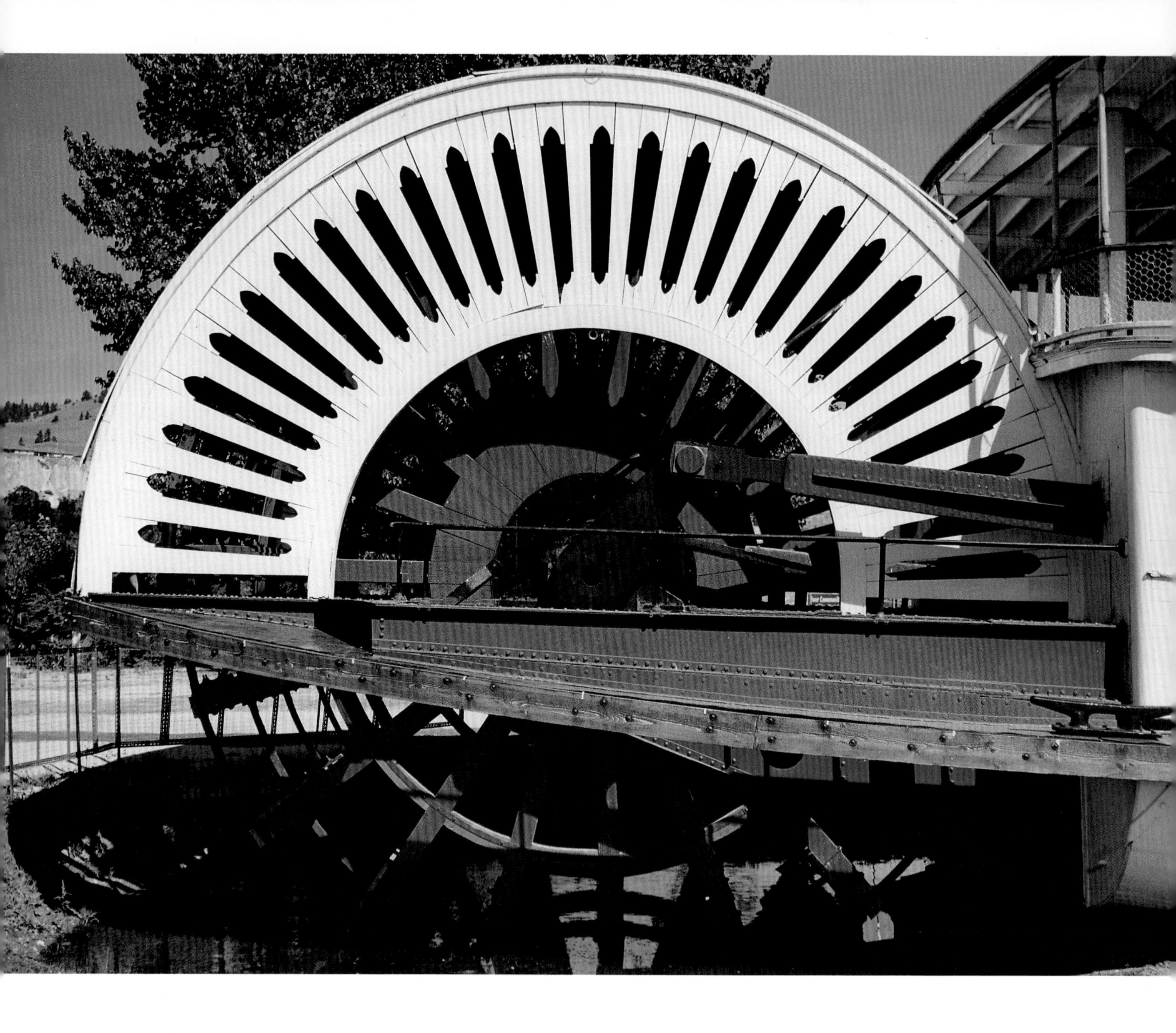

Retirement, Preservation and Restoration

The huge paddlewheel of the *Sicamous*, left; inside the paddlebox, above.
—ROBERT D. TURNER

The tug *Okanagan* was the last CPR vessel in service on Okanagan Lake and operated until 1972.—MAURICE CHANDLER

For over a decade the *Sicamous* floated quietly at Okanagan Landing. Through the Second World War and into the busy post-war years there was little to attract attention to the steamer and the once busy decks were silent except for the lapping of waves against the hull. It was clear that the *Sicamous* was not to run again for the CPR. In 1947 reports were published that the steamer was to be sold to the Yellowknife Transportation Company for use in the Northwest Territories but the proposal came to nothing. Other ideas were suggested but in the end the company was not prepared to sell the vessel for uses that would have seen it used, as Kelowna's *Courier* noted, "as a dance hall or some mundane commercial enterprise in her old age." The CPR preferred to see the steamer preserved as a historical landmark. Fortunately there were individuals who had similar ideas. The Penticton Gyro Club took on the project to use the vessel as a meeting place and community centre although the city of Penticton would be the owner.

On June 24, 1949, the CPR sold the *Sicamous* to Penticton for $1.00 but arrangements had to be completed for a permanent home. Finally on August 27, 1951, the steamer was towed to Penticton. Hundreds of people turned out in a welcoming celebration. The *Sicamous* was eased into a trench at the western end of the city's famous beach and then sand was backfilled.

In November, at the club's installation banquet, the key to the pilothouse was turned over to the new Gyro President, A. H. "Barney" Bent, by Mayor W. A. Rathbun. The Gyro Club then formally took over operation with a nominal lease from the city. Herb LeRoy became past president taking over from Dr. Herb McGregor, both of whom had been important figures in acquiring the *Sicamous*. Saving the *Sicamous* at Penticton was a considerable accomplishment particularly at a time when heritage preservation was in its infancy in British Columbia. After the end of her steaming days, the *Sicamous* lost numerous components. The main steam pipe that connected the boiler to the engines was taken out, and nearly all small fittings and the electric generator were removed. All furnishings were used elsewhere, sold or disappeared. A newspaper report called her a "dignified hulk," with few traces, except the sideboard or buffet, of her former grandeur.

The Gyro Club was faced with a challenging task. Just providing light, running water, washrooms, a kitchen and other amenities cost $18,000 in 1952. Structural alterations included removing the boiler tubes, the addition of a staircase from the freight deck to the saloon deck and the dismantling of the overnight cabins around the dining room to enlarge the lounge area. To support the upper deck in place of the stateroom partitions, mahogany posts, acquired in Nelson from the *Nasookin*, were substituted. The cambered deck in the dining room was covered over; a flat floor and stage were installed for dancing. Painting and cleaning were needed throughout the *Sicamous.* Opening day was May 24, 1952, preceded the night before by a party on board. Alderman Joe Harris acted for the city to open the vessel to the public. The Gyros also landscaped and developed the park area around the *Sicamous.*

In the years following, the *Sicamous* hosted many meetings, gatherings, weddings, dances, and special events. It became a well-known landmark of Penticton's waterfront. In the late 1950s, Penticton's museum was moved on board where it remained until the opening of a new building in July 1965. During this time the city took over operations and leased space to a series of businesses including a wax museum whose exhibits ranged from an alligator swamp to the Last Supper. A restaurant took over the saloon deck and undertook further modifications. Over the years several restaurants, of varying quality, operated on the *Sicamous.* They kept the *Sicamous* occupied, helped to pay maintenance costs, and many people enjoyed dining on board. However, they also took a toll. Restaurant development turned the ladies' saloon into a kitchen, in the process cutting two large ventilators into the elaborately curved ceiling, and the once bright dining room, including even the coloured, patterned glass clerestory windows, was painted brown. As well, a beautiful, etched glass door into the dining saloon disappeared.

By the mid-1980s, the *Sicamous* was clearly in peril. On the outer decks decades of temporary repairs and patches began to show their effects, and, often, lack of results. The pilothouse was devoid of fittings and the walls deeply carved with initials and graffiti. Major work was needed or it would eventually be lost through structural deterioration or, worse still, fire.

In 1988 the SS Sicamous Restoration Society was formed to undertake the rehabilitation and restoration of the *Sicamous.* With the co-operation of the city, the society set about planning and fundraising. A major step was the decision by the city not to extend the lease for the restaurant operating on the vessel. Its removal was essential to a successful rejuvenation of the *Sicamous*

For well over a decade, the *Sicamous* sat quietly at the Okanagan Landing shipyard, largely forgotten.—ED SCHIMPF

"Herb LeRoy got the notion to bring the *Sicamous* down here and the Gyro Club backed the idea and we purchased it from the CPR for a dollar. . . . We got permission from the city to beach it at its present site and we utilized a large bulldozer to move enough sand to back it in. Then the bulldozer pulled and pushed and bunted it into place. The hull was sound, most of the machinery had been removed from the engine room. The upper structure was pretty much as it is now, but it was in quite a state of disrepair. So, the Gyro Club took it on as a project for the community and patched it up and spent a lot of hard work . . . and funds to try to rehabilitate it. And we used to use it for meetings and banquets. . . . It served as a focal point for community activities and social events; there were lots of parties and dances, reunions and things of that nature."—DR. HERB MCGREGOR, FORMER GYRO CLUB PRESIDENT, 1994

In June 1951 the preparations of the Gyro Club were in hand and the *Sicamous* was towed to Penticton by the tug *Okanagan* under the command of Captain Walter Spiller. A trench was excavated and the big sternwheeler was eased back into the temporary lagoon that was then filled in and landscaped. —HERITAGE PHOTO CO-OP

The dining saloon, redecorated with the clerestory windows painted over, in use as a restaurant in the 1980s. Originally, there were staterooms along both sides of the saloon.—ROBERT D. TURNER

as a community centre. Eventually a five-phase plan was put into effect, which systematically outlined the restoration of many parts of the *Sicamous*, installation of fire-suppression systems, improved heating and utilities, its opening as a heritage vessel and community centre and, eventually, the restoration of the upper deck cabins.

Much of the work was dirty, hard and uninspiring. Ports into the hull had been left open and the bottom was filled with sand and debris, which had received quantities of water from disconnected drains and sewage. The cleanup was a major and unpleasant task. So much material was removed that during high water, the *Sicamous* briefly began to float. Old decking had to be removed, many layers of paint scraped, and a sprinkler system installed. Slowly the benefits of the work became apparent as the hull took on a like-new appearance. From under the brown paint of the dining room and the grease stains in the kitchen, gold leaf began to emerge, and the clerestory windows once again transmitted bright, sparkling Okanagan sunshine into the saloon. To make the *Sicamous* more functional, most of the outer walls were insulated and an inner sheeting of tongue-and-groove siding was installed to provide a traditional appearance. A further requirement was an upgraded heating system for the main public and office areas.

Visitors were welcomed for tours during the many months of work. Funding came from private and corporate donations, casino nights, federal employment and retraining programs, provincial community tourism employment programs, annual grants from the City of Penticton and a major grant under the B.C. Historic Landmarks Program. After thousands of hours of work, which included the installation of new hardwood flooring in the dining saloon, the *Sicamous* was opened for receptions, weddings, meetings and other social and business events in 1993. Much remained to be done to both the interior and exterior. By that point nearly 150 people had worked on the restoration, receiving job training and valuable work experience.

The retirement years of the *Naramata* presented a different story. After her last use, the boiler was drained, the engine was carefully lubricated and all of the steam lines cleared of moisture to prevent damage from freezing. Saved from vandalism by obscurity and being accessible only from the water, the *Naramata* remained in overall good condition.

In 1972 the shipyard site at Okanagan Landing was dedicated as Paddlewheel Park and several buildings saved for community use. To complement this project, the Okanagan Landing Association arranged with David and

James Keffer, the owners of the *Naramata,* to bring her back to the Landing. Eventually ownership passed to the association, and local Sea Scouts assisted with maintenance. However, by the mid-1980s, the scouts were no longer able to maintain the *Naramata.*

Silently, the tug floated offshore at the Landing. The most frequent visitors became pigeons that roosted in the cabins. By the late 1980s the *Naramata* was increasingly in need of attention. A proposal was developed to move the tug ashore, beaching it in a cradle at the Landing. However, the idea proved impractical. As interest in the vessel waned, it was proposed to sink her as a site for divers. Fortunately, Randy Manuel, director of the Penticton Museum, heard about the proposal and was invited to a meeting in Vernon in November 1990. "There was a unanimous vote to sell the ship to me for one dollar. I agreed, signed the papers, and drove home that night in a blinding snowstorm wondering what the hell stupid thing I had gotten myself into. I had no authority to buy the ship for the city or any other group. Yet I didn't want to see the vessel put to the bottom of the lake." The Kettle Valley Railway Heritage Society, which had been formed to try to preserve features of the KVR, "after a noisy meeting . . . bought the ship," and took on the task of restoring the tug and finding a permanent home. "Negotiations with the City of Penticton," Randy recalls, "went without a hitch."

On October 1, 1991, *Naramata* was towed from Okanagan Landing to Penticton. Berthed off the entry to the Okanagan River, near the *Sicamous,* the *Naramata* looked entirely at home. A cleanup began that included the removal of the pigeon droppings and a substantial quantity of material that had built up in the boat over the years. Unfortunately, during the fall of 1993, a very serious situation was discovered. The *Naramata*'s steel hull, at first thought to be in excellent condition, was actually paper thin in places and was leaking. The likely cause was that sulphur from the coal had produced a weak acid that had gradually corroded the hull from the inside. The situation was serious enough that if left to deteriorate the tug could have flooded and sunk. Repairs would have required a drydocking and replacement of some of the hull plating. With this work far beyond the resources available, it was decided to excavate a berth, and pull the tug onto the beach. Funded by a special grant from the provincial Heritage Trust, the *Naramata* was safely berthed and sand backfilled around the hull. While the possibility of one day refloating the tug was not precluded, work could proceed on refitting the cabins, decks and all of the equipment on board.

Marlene Pugh and Randy Manuel on the forward staircase, above left, which remains a highlight of the *Sicamous*. Intricate woodwork was a feature of the CPR sternwheelers: a ventilation panel over a stateroom door, above, and the ceiling of the ladies' saloon during restoration. *Naramata*'s compound vertical engine, below left, contrasts with the horizontal tandem-compound engines of the *Sicamous*, shown below, being inspected by Terri Reim and Marlene Pugh. Fred Tayler points out details of the starboard pitman arm and valve gear.
—ALL ROBERT D. TURNER

Reflections

The *Naramata* is the only Interior steam tug preserved in the province. Along with the coastal steam tug *Master*, based at Vancouver, they are the only tugs of the steam era, not rebuilt to diesel power, surviving in British Columbia. Like the few sternwheelers that have been preserved, they are a legacy for the future. As well as preserving an important technology, the *Naramata*'s heritage is a story of hard work and day-to-day activity that provided a critical connection directly or indirectly for thousands of people in fruit ranching, farming and in the packing industry throughout the Okanagan. Moreover, people living thousands of miles distant benefited from her presence by the availability of fruit and other products from the Okanagan.

The *Sicamous* began her career by providing an important, often critical, service to the communities of the Okanagan. In her most recent role the steamer continues to have a functional as well as symbolic place in the Okanagan. In fact, she has spent more time as a landmark at Penticton than she did in active CPR service and with the passage of a few years, more people will have attended functions, toured the vessel or had family celebrations on the *Sicamous* since her retirement by the CPR than travelled on her when she was an operating vessel.

The *Sicamous* represents the culmination of sternwheeler design and construction in western Canada and was the last built of the largest type of vessels operated in the Interior of British Columbia. The *Sicamous* was the ultimate extension of designs and service standards that, unknown to their creators, were more a reflection of times past and the optimism and tremendous expansion and investment of the early 1900s than they were of the decades to come. The Okanagan underwent immense growth in the decade before the First World War that saw a dramatic expansion of orchard lands and farms that would take some years to mature. The *Sicamous* was part of the transportation system that the CPR developed for the anticipated traffic in the years ahead. In those pre-war years, across its vast system, the CPR expanded its capacity, extended services and made great improvements. In many ways, the *Sicamous* represented the Canadian Pacific Railway in the Okanagan; it was a showpiece of a proud, successful company that valued its reputation and influence. The *Sicamous* was symbolic of the CPR's system-

Remembering . . .

"One summer when they had a bumper peach crop, I was on the *Okanagan*, we hauled 36 cars a day, Penticton to Kelowna, to go by rail out of Kelowna and the *Naramata* used to haul 16 cars out of Penticton then it would come back to Summerland to pick up a full load between Summerland and Naramata. Then the other tug the *Kelowna*, it handled Westbank, Peachland and Greata Ranch and Okanagan Centre, it did all those. One barge. It was busy around Kelowna, I'll tell you."

—ROBERT MANNING, FIRST OFFICER, 1994

"Some of the nice days, back in the 50s in the evenings. . . . Steam is quiet. We would be coming up the lake after supper. We'd get out on the barge and just lie out there and read and relax; peaceful. In the evening you'd watch the stars, the whole canopy of heaven would open up. It was beautiful. We had our games of cards, chess, chequers, pocket books. The most exciting times were in electrical storms. We had the old wooden barges with the hog chains and hog posts to stiffen the barge. The hog chains were made of metal. In an electrical storm it was nice to watch them. Bizzzzzz! Right to the water. Scary; there is quite a bang to it. You'd watch the thunder cloud roll away down the lake."

—BILL GUTTRIDGE, FIRST OFFICER, 1994

"None of us had any realization that it would be so short, that it would be taken off the passenger service in 1935. I always regretted that we didn't once go all the way again [to the Landing]."

—BONNIE DAFOE, 1993

"The crew would be back and forth for coffee. You'd have potatoes, two vegetables and two kinds of meat for lunch. For supper, you'd have either fish or liver or whatever you thought they would like and potatoes. You always had potatoes. And two kinds of vegetables for every meal. Desserts. They often didn't eat dessert. I'd make two or three pies in the morning and sometimes there would be some left the next day."
—MARY MERRIFIELD, RELIEF COOK, 1994

"Today we are inclined to think that nothing is good unless it is streamlined, but we also look to the past for guidance in many things. The Gyro Club thinks that it is a worthwhile thing we are doing in preserving this vessel."
—BARNEY BENT, GYRO PRESIDENT AND ONE OF THE FOUNDERS OF THE PIONEERING AVIATION FIRM OKANAGAN HELICOPTERS, QUOTED IN *The Penticton Herald*, MAY 29, 1952

Naramata and *Sicamous* at Penticton, 1995.—ROBERT D. TURNER

wide standards for service and accommodation in its trains, hotels and steamships. Her roots go back through other important and pioneering vessels that came from the designs of James W. Troup, John C. Gore, Thomas, David and James Bulger, Dave Stevens, George Keys and others in a lineage that began on the lower Columbia in the mid-1800s. Moreover, in her design are preserved features that Troup also incorporated into the construction of the famous *Princess* steamships of the CPR's British Columbia Coastal Steamship Service; not one of the early *Princesses* has survived. The *Sicamous* represents far more than just one vessel but an evolution of inland and coastal steamship architecture and technology that spans several major river systems, coastal shipping routes from Portland to Alaska and two countries for at least half a century.

The First World War, with its many impacts, the extension of the railways through the Okanagan and nearby districts, and, most importantly, the growing versatility and popularity of the automobile, doomed steamer travel in the Okanagan. Although by the mid-1920s, few people travelled on the *Sicamous*, it still carried large volumes of high-value fruit during the peak summer and early fall season, as well as express and mail. The Great Depression of the 1930s was the final blow. The sternwheelers were catalysts in decades of great change in the Okanagan and they were of immense importance.

The *Naramata* reflected different ideas and needs. Intended to provide a link in the railway system for carload freight and fruit shipments, it would remain viable for over half a century before it also gave way to highways, motor vehicles and the changing nature of the agricultural industry.

As the years go by fewer and fewer people will remember the steaming days of either the *Sicamous* or the *Naramata* as part of the Canadian Pacific's British Columbia Lake & River Service. However, many more will become acquainted with the two steamers in their continuing roles as heritage vessels and community facilities. In the years ahead, they will continue to have an important impact on the Okanagan in ways that their builders could never have imagined: as heritage centres providing a sense of time and place to residents and visitors and as economic generators for the Okanagan. The *Sicamous* was built in large measure to encourage tourism and travel to the Okanagan in the early 1900s. This fundamental role for the elegant sternwheeler, emerging from its restoration over 80 years later, in fact, has changed very little.

Romantic artwork was often a feature of Canadian Pacific's advertising and this brochure from 1919 for the "Lake District" is a charming example. It promoted CPR services and hotels and the Okanagan and Kootenays as holiday destinations. The steamer, with some artistic license, resembles the *Kuskanook* and *Okanagan*, although it actually appears to have been based on the Great Northern's *Kaslo*.
—ROBERT W. PARKINSON COLLECTION

Preserving the *Sicamous* and the *Naramata* is an ongoing project and we welcome your help. We invite you to join the SS Sicamous Society; donations and support are always needed. Artifacts, including fixtures and furnishings from the vessels, will help complete the restoration work. Photographs and personal recollections are also needed. Please join with us in this rewarding heritage program.

Further Reading

Edward Affleck. 1973. *Sternwheelers, Sandbars & Switchbacks.* The Alexander Nicolls Press, Vancouver, B.C.

Art Downs. 1992. *British Columbia-Yukon Sternwheel Days.* Heritage House Publishing Co., Surrey, B.C. (Originally published in 1972 as *Paddlewheels on the Frontier*).

Barrie Sanford. 1977. *McCulloch's Wonder. The Story of the Kettle Valley Railway.* Whitecap Books, West Vancouver, B.C.

Robert D. Turner. 1984. *Sternwheelers & Steam Tugs. An Illustrated History of the Canadian Pacific's British Columbia Lake & River Service.* Sono Nis Press, Victoria, B.C.

Robert D. Turner. 1991. *The SS Moyie, Memories of the Oldest Sternwheeler.* Sono Nis Press (and the Kootenay Lake Historical Society), Victoria, B.C.

This book is a companion volume to *The Sicamous & The Naramata.*

Robert Turner has spent many years researching western transportation history. Formerly Chief of Historical Collections, he is now a Curator Emeritus at the Royal British Columbia Museum and is also Project Historian for the SS *Moyie* National Historic Site. *The Sicamous & The Naramata* is his eleventh book.